The
Internet for Cell and Molecular Biologists
Current Applications and
Future Potential

Edited by:

Andrea Cabibbo (Ph.D.)
Dept. of Biology, University of Rome
Tor Vergata, Via della Ricerca Scientifica snc
00133 Rome, Italy

Richard P. Grant (D.Phil.)
Division of Structural Studies
MRC Laboratory of Molecular Biology
Hills Road, Cambridge, CB2 2QH, UK

Manuela Helmer-Citterich
Centre for Molecular Bioinformatics
Dept. of Biology, University of Rome
Tor Vergata, Via della Ricerca Scientifica snc
00133 Rome, Italy

horizon press

Copyright © 2002
Horizon Scientific Press
32 Hewitts Lane
Wymondham
Norfolk NR18 0JA
U.K.

www.horizonpress.com

British Library Cataloguing-in-Publication Data

A catalogue record for this book is available from the British Library

ISBN: 1-898486-32-8

Distributed exclusively in the United States, its dependent territories, Canada, Mexico, Central and South America, and the Caribbean by Springer-Verlag New York Inc, 175 Fifth Avenue, New York, USA, by arrangement with BIOS Scientific Publishers Ltd, 9 Newtec Place, Magdalen Road, Oxford OX4 1RE, UK.

Distributed exclusively in the rest of the world by BIOS Scientific Publishers Ltd, 9 Newtec Place, Magdalen Road, Oxford OX4 1RE, UK.

Printed and bound in Great Britain

Contents

Contributors

Andrea Cabibbo
Dept. of Biology
University of Rome
Tor Vergata
Via della Ricerca Scientifica snc
00133 Rome
Italy
andrea.cabibbo@uniroma2.it

Barbara Brannetti
Centre for Molecular Bioinformatics
Dept. of Biology
University of Rome
Tor Vergata
Via della Ricerca Scientifica snc
00133 Rome
Italy
brannett@uniroma2.it

Lorenzo M. Catucci
Centro di Calcolo e Documentazione
University of Rome
Tor Vergata
Via della Ricerca Scientifica snc
00133 Rome
Italy
lorenzo@sancho.ccd.uniroma2.it

Fabrizio Ferré
Centre for Molecular Bioinformatics
Dept. of Biology
University of Rome
Tor Vergata
Via della Ricerca Scientifica snc
00133 Rome
Italy
fabrizio@cbm.uniroma2.it

Richard P. Grant
Division of Structural Studies
MRC Laboratory of Molecular
Biology
Hills Road
Cambridge
CB2 2QH
rpg@mrc-lmb.cam.ac.uk

Manuela Helmer-Citterich
Centre for Molecular Bioinformatics
Dept. of Biology
University of Rome
Tor Vergata
Via della Ricerca Scientifica snc
00133 Rome
Italy
citterich@uniroma2.it

Michele Quondam
Dept. of Biology
University of Rome
Tor Vergata
Via della Ricerca Scientifica snc
00133 Rome
Italy
micheleq@libero.it

Allegra Via
Centre for Molecular Bioinformatics,
Dept. of Biology
University of Rome
Tor Vergata
Via della Ricerca Scientifica snc
00133 Rome
Italy
allegra@cbm.bio.uniroma2.it

Preface

The main message of this book is that a personal computer is nowdays a terminal to an exponentially increasing amount of very relevant and completely free information. 'The world in a box' is now a reality. For the exploration of the 'Bio-Web' proposed in this book, a decent computer (see Chapter 2) connected to the internet will suffice. Any operating system will do. The only required software is a web browser. First, a general and practical framework for the understanding and non-trivial usage of computers, web sites, e-mail, newsgroups, search engines, PubMed and more is provided (Chapters 1-5, 11-13). Second, a number of more strictly biological issues, mainly related to the analysis of biological sequences and structures, are dealt with in detail (Chapters 6-10).

The book is targeted at a wide audience of students, post-docs, researchers, professors and educators. It also will be useful for those generally interested in biological or genetic issues, who wish to acquire a tighter handle on the usage of computers and the internet in biology. A more expert audience will find this book a useful reference, as it contains a wealth of well organised and annotated pointers to resources in many areas of biology and bioinformatics.

All chapters are quite self-contained and can be read in the preferred order. Chapters 6 to 9 constitute a 'core' on DNA and protein sequence analysis that might be better read as a unit. In particular, chapter 7 illustrates general concepts about sequence alignments that are widely used in the following two chapters.

Feedback to the authors through e-mail is warmly encouraged.

Enjoy your reading!

Andrea Cabibbo Ph.D.
andrea.cabibbo@uniroma2.it

Books of Related Interest

Gene Cloning and Analysis: Current Innovations

Genetic Engineering with PCR

An Introduction to Molecular Biology

Probiotics: A Critical Review

Prions: Molecular and Cellular Biology

Peptide Nucleic Acids: Protocols and Applications

Intracellular Ribozyme Applications: Principles and Protocols

Prokaryotic Nitrogen Fixation: A Model System for the Analysis
   of a Biological Process

Molecular Marine Microbiology

NMR in Microbiology: Theory and Applications

Oral Bacterial Ecology: the Molecular Basis

Development of Novel Antimicrobial Agents: Emerging Strategies

Cold Shock Response and Adaptation

Flow Cytometry for Research Scientists: Principles and Applications

Helicobacter pylori: Molecular and Cellular Biology

The Spirochetes: Molecular and Cellular Biology

Environmental Molecular Microbiology: Protocols and Applications

Advanced Topics in Molecular Biology

Industrial and Environmental Biotechnology

Genomes and Databases on the Internet

Microbial Multidrug Efflux

Emerging Strategies in the Fight Against Meningitis:
   Molecular and Cellular Aspects

The Bacterial Phosphotransferase System

For further information on these books contact:

Horizon Scientific Press, 32 Hewitts Ln, Wymondham, Norfolk, NR18 0JA, U.K.
Tel: +44(0)1953-601106. Fax: +44(0)1953-603068. rab@horizonpress.com

Our Web site has details of all our books including full chapter abstracts, book reviews, and ordering information:

www.horizonpress.com

1

The Internet:
All You Wanted to Know
and Didn't Dare to Ask

Lorenzo M. Catucci and Manuela Helmer-Citterich

Contents

Abstract

In the last 10 - 15 years the computer became an essential companion for cell and molecular biologists. At first personal computers were mainly used as word processors or to produce nice pictures for papers or talks. In many research institutes mainframes were set up as mail servers and to host and run the first packages of accessible bioinformatic tools: the Staden (http://www.mrc-lmb.cam.ac.uk/pubseq/staden_home.html; Staden *et al.*, 2000), Intelligenetics (Intelligenetics Suite, Intelligenetics, Inc. Mountain View. CA) and GCG (now at http://www.accelrys.com/about/gcg.html) packages. Sequence databases were slowly

From: *The Internet for Cell and Molecular Biologists: Current Applications and Future Potential*
ISBN 1-898486-32-8 © 2002 Horizon Scientific Press, Wymondham, UK

starting to grow. No, or very little, organized information about the new tools was available in the academy, but a lot of know-how was passing hand to hand in the research labs.

Since then, things have changed a lot. Each personal computer is now much more powerful and flexible than those old mainframes. Many new and sophisticated tools were developed to help biologists in their work and, most importantly maybe, many tools for advanced communication became commonplace. This had a very strong impact on the experimental biologist's life.

Every computer, if equipped with an ethernet card (or a modem) and an internet connection, represents a node in an immense network. It becomes therefore a window to the outside world and the outside world offers a lot of interesting information: from Medline access to web pages dedicated to specific hot topics of biological interest. There is no hope of giving an exhaustive list of all the useful and interesting places that can be visited during an internet trip. It is always worth trying to look around, to bookmark new sites and explore different tools.

Almost every biologist has experience in the use of electronic mail and internet browsers, but sometimes feels not completely at ease with the matter. Very few biologists received a well-organized instruction in the use of informatics instruments for biology, so what to do when we need something more, such as choosing the best parameters in a sequence search and understanding all the implications of a complicated and sometimes almost unreadable output file? We look around and find nice web pages, full of information and we would want to be able to design our own web site: do we really need to ask someone else for help? We want to visualize a protein structure on the screen or try to understand the possible consequences of a residue mutation: can we afford to play with molecular graphics?

This manual was designed as a sort of 'cook book' to try to fill some of the gaps that may affect the life of a biologist who missed an organized preparation in basic informatics, but still wants to be able to take advantage of skilful use of computers and of the rich internet tool-set.

Let us start with a short description of the basic elements in computer networking.

1.1 A network of nodes

Each computer can be a node of the network (Tanenbaum, 1996) also known as the internet and therefore it needs a unique address to be identified in the network. It needs also a set of rules (generally called 'protocols') to exchange data with the other nodes of the network. Actually the most used protocol is known as TCP/IP, which stands for Transmission Control Protocol/Internet Protocol.

You don't need to know much about TCP/IP, since you will probably find it already installed in your personal computer as part of your OS (Operating System), and only three pieces of information are required:

1.   The IP address assigned to your personal computer (Figure 1.1 for Macintosh and Figure 1.2 for Windows).
2.   The subnet mask enables systems to automatically discern

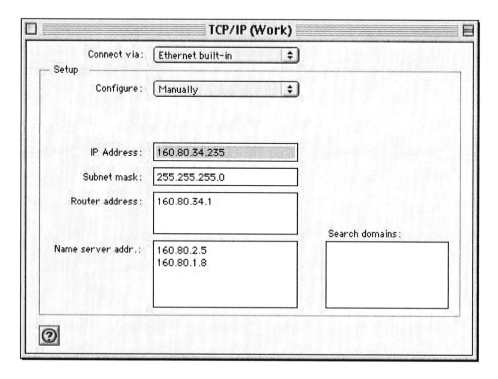

Figure 1.1. The IP address assigned to your personal computer (Macintosh).

IP addresses belonging to the same LAN (whose traffic is not to be routed) from the external ones.

3.   The IP address of the router machine that connects this LAN to the rest of the world (Figure 1.1 or Figure 1.2).

An example of an IP address is 160.80.34.242, which corresponds to the server of our research group. The first three bytes identify the department (160.80.34 is common to every computer in the department), while the last byte at the right identifies the single node (one computer). We will not describe here the role and meaning of the subnet mask and of the router, since this would be out of the scope of this chapter; these numbers are usually identical for all the nodes in the same LAN.

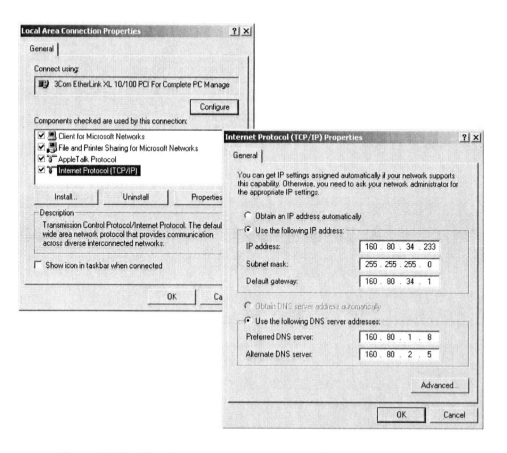

Figure 1.2. The IP address assigned to your personal computer (Windows).

Each computer can also be identified by a name, usually in the form: computer_name.department_name.something.something. Our server, for instance, is identified by the name cbm.bio.uniroma2.it: **cbm** is the name of the computer (Centre for Molecular Bioinformatics), **bio** identifies the Department of Biology, which is part of the University of Rome Tor Vergata (**uniroma2**); at the end of the name **it** stands for Italy. The last characters of a computer name can identify the country (**uk** for Great Britain, **fr** for France and so on) or an academic site (**edu**), an organization (**org**), government (**gov**), military (**mil**) or a commercial site (**com**). See also **domain names** in the short glossary reported below.

You can select a computer both by its IP number and by its name. If you use the number, and the other computer is on, contact is made directly. If you use the name (that for some people is easier to remember), the protocol accesses a table of correspondence between numbers and names and then establishes the contact. The distributed hierarchical database of the correspondences is called the Domain Name System (or DNS). This is the reason why you must specify the IP address of a nearby DNS Server, in order to be able to use names (in addition to IP numbers) to identify other nodes in the internet network (Figure 1.1 or Figure 1.2). In some cases you can define the search domain, i.e. the domain name that the TCP/IP automatically searches when you do not write the whole name of a computer.

When you specify the name of a computer to establish an internet connection (see below), the DNS is contacted through the TCP/IP, the name is transformed in an IP number and the connection is set up.

1.2 Network services

Internet users may exploit the following services:

- Electronic mail (e-mail): permits you to send and receive mail (see Chapter 12)
- Discussion groups: groups where different topics are discussed through a distributed system (see Chapter 11).
- Remote shell access: permits you to log into another computer and use it as if you were there.

- FTP (File Transfer Protocol): allows your computer to rapidly exchange files with a remote computer (see Chapters 4 and 12).
- World Wide Web (WWW or "the Web"): the largest, fastest growing activity on the Internet.

With the WWW, you can retrieve texts and documents, view images, animation, video, listen to sound files, speak and hear voice. You can also execute remotely or download programs, as described in the next chapters. The WWW is now mostly based on special documents written in the HTML (Hyper Text Markup Language) language. Hypertext pages contain links, which are special 'clickable' words or areas in a page, which trigger a transition, such as the link to another web page, the starting of a download, the execution of a program and so on.

Here's essentially what happens when you type a URL (Universal Resource Locator) in the location box of your browser:

- You write the web address of the computer you want to contact (for instance the EBI web server www.ebi.ac.uk, in Britain), preceded by **http://** (Hyper Text Transfer Protocol) and followed by more words usually indicating WHERE in the remote server the interesting data are stored. These words may represent folder or file names and are separated by slashes ('/'). A web address such as: http://www.ebi.ac.uk/Tools/index.html, for instance, means that you contact the www.ebi.ac.uk server and you browse the index.html file that is found in a 'Tools' directory (a directory is the Unix equivalent of a folder in the Windows or Mac systems).
- Your browser reads the html file (index.html) in the indicated directory (Tools) and displays its content (eventually with graphics) on your screen.

In a fast and easy way, you downloaded information (text and figures) from a remote computer. The information is there and you can access the list of tools offered by the European Bioinformatics Institute. And much more: you are also allowed to use the software tools via web, using the time and resources of a remote computer.

A modern biologist must be able to deal with many different tools: micro-pipettes, precision balances, petri dishes, centrifuges of different sizes, PCR machines, micro-array tools and, last but not least, internet and computers.

1.3 Short glossary

Alphanumeric – generic term indicating alphabetic characters (A-Z) and numbers (0-9).

Algorithm – defines a procedure to solve a problem through a series of ordered and finite elementary steps. It may be represented schematically as a flow-chart. Programming languages are tools for implementing algorithms.

Applet – a small program written in the Java language that can be executed in an HTML page.

ASCII (American Standard Code for Information Interchange) – standard code for data exchange. It codes up to 128 (2^7) different symbols: 96 alphanumeric characters and 32 control characters (for peripherals and other communication tools).

BIT (BInary digiT) – elementary information unit of the binary system: it can assume only two values [0] and [1], corresponding to the electronic status [off] and [on]. 8 bits constitute a Byte, that can be used for representing characters, such as letters [A-Z] and numbers [0-9].

Browser – a 'client' program that is used to access documents via the Internet. Browsers can be both text-based (Lynx, Gopher) or graphical (Netscape, Internet Explorer, Opera).

Byte – information unit composed of 8 bits, utilized to represent alphanumeric characters in the computer processor. The byte may assume 256 (2^8) different values, including zero.

Clock – electronic device generating regular pulses in order to synchronize processor operations. The clock speed is measured in MHz and is related to the computer power (cf. Chapter 2).

C language – programming language developed by D. Ritchie and K. Thompson in 1972/73 in the Bell Laboratories (USA). The C language is quite difficult (when compared with other programming languages more similar to natural languages), but is very effective with complex problems.

Client – a remote computer (or software) used to connect and obtain data from a server computer (or server program). A web browser is a client software able to communicate with a remote host running the appropriate web server.

Cookie – represents a small piece of information sent by a Web Server to a Web Browser, which is stored, for a defined length of time, in the client computer. The Browser may accept or not accept the Cookie, according to the user preferences. Cookies generally contain information such as login or registration information, user preferences, etc.

Bookmark – frequently accessed links can be saved in a bookmark file.

DATABASE – is a big ensemble of data all organized on a magnetic medium (usually a hard disk) or a CDROM and structured for the optimization of search and retrieval, updating and printing reports.

Default – indicates the parameter value automatically chosen by the software (or by the OS) when not explicitly specified by the user.

Domain name – the address identifying an internet site and generally composed of at least two parts separated by dots. The name on the left is the more specific one. All the machines in a LAN share the right-hand portion of their domain names.

DNS – Domain Name System; a database system that translates an IP address into a domain name.

E-mail – Electronic Mail: messages, usually text, but also graphics, sent from one person to another via computer.

Ethernet – a common physical mean of connecting computers in a LAN; it is used to distribute data, video and sound signals in a network connecting up to 1024 personal computers at very high speed.

FAQ – Frequently Asked Questions; a list of the most common questions posted on a newsgroup.

File – Ordered ensemble of data that can be memorized or elaborated on an electronic device. Every file must have a name, according to rules that are OS-specific.

Firewall – A combination of hardware and software that separates a LAN into two or more parts for security purposes.

FORTRAN (FORmula TRANslation) – Programming language oriented to mathematics and scientific research.

Freeware – Free software available on the Internet that can be redistributed.

FTP (File Transfer Protocol) – A common protocol for moving files across two internet sites. With FTP you can access a remote computer and send or download files. Many internet sites host public file archives that can be downloaded freely using the username *anonymous* and your email address as password. These sites are called anonymous ftp servers.

Gigabyte – equivalent to 2^{30} bytes, 1000 or 1024 Megabytes, depending on who is measuring.

Hardware – all the physical parts composing the computer: chip, cards, memory, peripherals…

Homepage – is the main page on a web site and usually acts as the starting point for navigation.

HTML (Hyper Text Markup Language) – The language used to create Hypertext documents for use on the World Wide Web (see Chapter 4)

http (HyperText Transfer Protocol) – HTTP is the most important protocol used in the World Wide Web (WWW) and is used for moving hypertext files across the Internet.

Hypertext – any text that contains links to other documents or to other words or phrases in the same document.

Internet – the world-wide inter-connected network of computers based on the TCP/IP protocols.

Intranet – A network inside a company or organization that uses the same kinds of software that are used in the Internet, but that is only for internal use.

IP Number (Internet Protocol Number) – a numeric address consisting of 4 parts separated by dots; it is translated into a domain name by the DNS.

Java – is a network-oriented programming language specifically designed for writing programs that can be included in web pages and executed through the Internet. Small Java programs are called 'Applets' and can offer functions such as animations, fancy graphics and so on.

Kilobyte - 1024 (2^{10}) bytes.

LAN (Local Area Network) – a network of computers sharing resources (hard disks, programs, databases, printers etc.) limited to the immediate area, usually the same building or floor.

Megabyte – equivalent to about one million (2^{20}) bytes.

MIME (Multipurpose Internet Mail Extensions) – an extension to the Internet mail protocol that allows non-text files, (such as graphics, programs, etc.) to be sent as attachments to regular e-mail messages.

Modem (MOdulator, DEModulator) – computers connected together over phone lines with a modem are able to exchange files, and to communicate with each other.

Navigate – to move in the internet by following hypertext paths, eventually jumping from document to document and from computer to computer.

Netiquette – the *etiquette* that guides online interaction on the Internet.

Network – a minimal computer network can be composed of two computers sharing resources, such as: hard disks, databases, printers and so on.

Newsgroup – the name for discussion groups on USENET (see Chapter 11).

Operating System – the sophisticated and complex program (composed of many programs and routines) which manages the user-computer interaction through simple actions or commands. Moreover, the operative system coordinates and controls the hardware performance and the peripheral usage. It usually resides on the hard disk and is started when the computer is started. Widespread operating systems are Windows, Macintosh, Unix, and Linux.

PIXEL (PIcture ELement) – elementary component of an image (composed of many pixels) represented by a dot on the display. The smaller the pixel, the more defined is the image.

Platform – the computer or operating system where a software application runs. Common platforms are Windows, Macintosh, Unix, and Linux.

Plug-in – an optional piece of software adding features to a larger piece of software. Browsers may host plug-ins for displaying files in different formats, such as: pdf files, or coordinate files (that can be visualized as three-dimensional molecules with appropriate plug-ins, see Chapter 10).

PPP (Point to Point Protocol) – protocol that allows a computer to use a regular telephone line and a modem to make TCP/IP connections.

Procedure – sequence of programs and paths organized for the solution of a specific problem.

Program – the same as software. Sequence of instructions written in a programming language (such as FORTRAN or C) with the aim of specifying to the computer how to operate on the input data in order to generate the output according to the programmer's need.

Programming language – artificial language used to write computer programs. A programming language, through precise rules and symbols, can be used to implement algorithms and describe actions and objects. Each programming language owns an alphabet (symbols with specific meaning), a grammar (rules for word recognition and to build correct sentences) and a semantic (rules for the assignment of a meaning to the sentences). Any program must anyway be translated in binary code to be recognized and executed by the machine. To this aim, special programs, known as *compilers*, have to be used.

RAM (Random Access Memory) – computer memory characterized by a very high speed in reading and writing. It is very important for determining the computer performances and its capacity is measured in bytes (cf. chapter 2).

ROM (Read Only Memory) – computer memory containing programs and data that cannot be modified by the user and are available only for reading. Its capacity is measured in bytes.

Router – special purpose computer that can connect a local network to the Internet. Routers look at the destination addresses of the packets passing through them and addressing them on a specific *route*.

Scripting language – high level languages can be used to specify computer procedures known as *scripts*. They do not need to be compiled (translated in binary code), but are executed by a special program called an *intepreter*. A script run is generally slower than the execution of a program of equivalent complexity.

Search engine – program on the Internet that allows users to search through massive databases of information. Very useful search engines can be used to search with keywords into indexed databases of web sites, allowing the identification and immediate link to sites containing information related to the specified keywords.

Shareware – software available for downloading on the Internet. Users who want to use the program are expected to pay a registration fee. In return they get documentation, technical support, and any updated versions.

SMTP (Simple Mail Transfer Protocol) – standard protocol for delivering e-mail.

Software – the ensemble of programs, documents, rules and procedures about data processing.

Software package – ensemble of programs and manuals dedicated to the solution of a specific problem.

SQL (Structured Query Language) – a specialized programming language for sending queries to databases.

Subnet mask – a number used to identify a sub-network so that an IP address can be shared on a LAN (Local Area Network).

TCP/IP [Transmission Control Protocol (TCP) and the Internet Protocol (IP)] – protocols that allow different types of computers to communicate with each other.

Terabyte – 1000 gigabytes.

URL (Universal Resource Locator) – address used to find a particular resource.

1.4 References

Staden, R., Judge, D.P. and Bonfield, J.K. 2000. Sequence assembly and finishing methods. In: Bioinformatics. A Practical Guide to the Analysis of Genes and Proteins. Second Edition. Eds. Andreas D. Baxevanis and B. F. Francis Ouellette. John Wiley and Sons, New York, NY, USA.

Tanenbaum, A.S. 1996. Computer Networks. Prentice-Hall Inc. Upper Saddle River, NJ.

2

Select the Right Computer

Michele Quondam

Contents

Abstract

This is a simple guide to the computer market: if you need a computer, you can now discover how to get the best and cheapest solution for your specific needs. This chapter also provides some information about computer components and their impact on the overall computer performance.

From: *The Internet for Cell and Molecular Biologists: Current Applications and Future Potential*
ISBN 1-898486-32-8 © 2002 Horizon Scientific Press, Wymondham, UK

An Apple, IBM compatible computer (also known as 'personal computer' or 'PC'), or Silicon Graphics computer share similar features and are very often built with the same components: they all have a central processing unit (CPU), a motherboard to control the CPU communications with other computer parts, a video board connected to a monitor, a hard disk to store the data, and some RAM (Random Access Memory) modules. Often other boards are mounted on the motherboard to add some specific features, such as a modem, a network card, a sound card, and many others. Some personal computers destined for heavy computational tasks or to run sophisticated graphical applications, can contain two CPUs (parallel machines) or have different and more complex architectures.

The parts that determine the computer speed and power are first of all the CPU type and speed, bus architecture and the quantity of available RAM, then the motherboard and the video card speed, and finally, the hard disk type and the presence of other peripherals.

2.1 CPU

The CPU is the 'brain' of the computer. The performance of the processor determines how quickly the computer responds to the requirements placed on it, and is best measured by counting instructions per second. Performance is often measured by clock speed, which refers to the number of pulses per second generated by an oscillator that sets the tempo for the processor, and is measured in megahertz (MHz) or gigahertz (GHz). However, some CPUs can process more than one instruction per 'tick' of the clock and therefore the clock speed is not appropriate for comparing the performance of chips from different manufacturers. There is no simplistic, universal relationship among clock speed, 'bus speed', and millions of instructions per second (MIPS).

Now (early 2002), the race to reach 2 GHz has started for IBM-PC systems, and the same is happening for 1 GHz CPUs for Apple computers. We can choose between CPU manufacturers for PC systems from among Intel, AMD and Cyrix, whereas CPUs for Apple systems come from either Motorola or IBM, and the consumer can not choose.

All brands have different processor types and each processor type runs at different speeds; in addition every 3-4 months a new processor arrives on the market. So, many different types of processors are available: since it would be very hard to describe them all, we can simplify the discussion by subdividing the CPUs into three groups:

Entry level CPUs (AMD Duron, Intel Celeron, and Apple G3); these chips are designed for inexpensive computers and are best suited for office applications (word processing, accounting), Internet access and light multimedia tasks (viewing video, digital photography, etc.). The most important feature of these different CPUs is the size of the cache. Cache is where the processor stores frequently accessed instructions or data for faster performance, a larger cache assures better performances. Cache sizes of entry level CPUs range from 128 Kb to 512Kb. Higher cache sizes are to be preferred.

Mid level CPUs (AMD Athlon, Intel Pentium III, and Apple G3 high); these chips provide a good computing power, suitable for handling intensive tasks like audio/video editing and 3D imaging as well as speeding up the rest of your computing tasks.

Top level CPUs (AMD Athlon 266 MHz bus, Intel Pentium 4 and Apple G4); these are the fastest processors available at the end of 2001. They are best suited for fast multimedia performance and heavy computational tasks.

2.2 Memory

The amount of available memory in a computer is very important, it is the place where the operating system, programs and data in current use are kept, ready to be accessed by the processor. If you are faced with a choice between buying a faster processor without much memory, or buying a slower processor with more memory you should select the second option. Extra RAM will provide more speed to the system than a few extra MHz of CPU speed.

Computer Memory amount is measured in megabytes (MB).

Any computer you are considering should have a minimum RAM of 128MB, preferably more. RAM is becoming increasingly cheap, so equipping a computer with 256 MB or 512MB of RAM is now much less expensive with respect to the last few years.

2.3 Hard disk

While RAM stores information for a limited period of time, the hard drive stores information permanently. The capacity of hard drives is measured in gigabytes. In new systems, they range from 20GB to 60GB in size, but they will probably reach 100-200 GB in the near future. A 20-30GB hard disk is a good choice for common use.

2.4 Video card

Also known as graphics cards, these components are responsible for displaying images on your monitor. The speed of the video card limits the overall computer speed. Several different types of video cards are available. It is therefore important to select the one that best suits your needs: high power is required by video games and graphical editing, while common office uses don't need great power. A video card has some memory modules. The memory indicates the video card capacity: with more memory a video card can attain higher display resolutions. At the moment the video card memories go up to 64MB.

2.5 Monitors

There are two types of monitors available for PCs: the traditional CRT and the newer LCD. The CRT ("*cathode ray tube*") is used for both televisions and computers. LCD ("*liquid crystal display*") monitors, also known as "*flat panel displays*", are used in notebook computers and more and more frequently for desktops as well. They are lighter and smaller than CRT monitors and feature reduced electromagnetic emissions and lower power consumption. Often a biologist must work several hours at the computer: buying a good CRT monitor (17" or 19" of size), or a LCD monitor (15" or 17") can be the best choice.

2.6 Other parts

Typically a computer is also equipped with a CD-ROM reader (or RW, reader and writer), an audio board for music and sound effects, a network card to communicate with other computers and access the internet.

2.7 The computer power and cost

Moore's Law, from the name of the Intel corporation president, states that the CPU transistor number doubles every eight - ten months. In other words, the CPU speed doubles every eight - ten months and the computer speed approximately every two years.

Despite this, the CPU price remains quite stable: in the last ten years the cost of the best CPU available in the market has always ranged around 150 - 200$.

This is probably the most important clue to the computer market: the CPU cost is determined by the number of sold CPUs and by their age: a cheap CPU will sell more than a more expensive one and will cost much less, but when a new CPU enters the market the older CPU will progressively leave it, and usually this is the cheaper.

So the market is in a sort of steady state: some older and cheaper CPUs, some middle level CPUs, and some high level CPUs: the market situation at the end of 2001 is shown in the Table 2.1.

Figure 2.1 shows the computer power growth in MHz CPU for the three processor categories.

This discussion about computer market holds generally true for any computer components, like hard disks, video boards, and others.

Table 2.1. The computer market in 2001.

Entry level CPU	Middle level CPU	High level CPU
Intel Celeron 733-800 Mhz	Intel P3 1000 Mhz	Intel P4 up to 1800 Mhz
AMD Duron 700-800 Mhz	AMD Athlon 800-1000 Mhz	AMD Athlon 1100-1400 Mhz
Mac G3 500 Mhz	Mac G3 600-800 Mhz	Mac G4 700-800 Mhz
50$	100-150	150$ and more

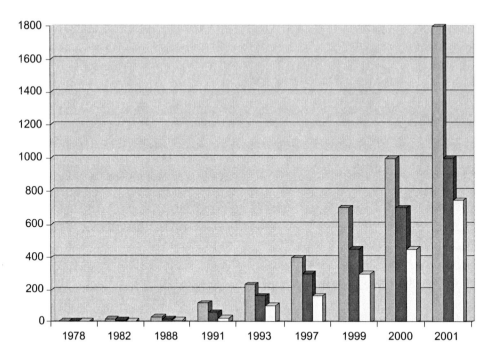

Figure 2.1. The computer power growth in MHz CPU for the three processor categories.
Grey=High level CPUs   Dark grey=Middle level CPUs White=Low level CPUs.

2.8 A computer to do what?

Consideraing the rapid evolution of computers, how can we select the
best option? We should ask ourselves the question: '*What do I have to
do with it?*'. Ordinary work, surfing the internet, playing that latest
hot game, or make a molecular simulation? Who will be using the
computer and what are his/her interests?

If we use the computer only for writing, internet access and
browsing we do not need an expensive computer; a few years old
computer might very well do the job. If we need a machine to develop
software, do some computer graphics and see video stream, we need
something better. Obviously other tasks can require more power: the
most power is needed to play the latest video-games, to make a movie,
to run 3D graphical applications and for other particularly computing-
intensive tasks.

For everyday usage, biologists need a machine to browse the

internet, write papers, and prepare graphs presentations. So the need for power is often limited; you can see on the figure (the white bar) that the cheaper computers today have the same power as the best two-year old computers: usually the cheaper computer among the newly available on the market represents the best compromise: the maximum power at the minimum price.

Our general advice is in any event the following: buying more memory or better quality monitors (i.e.: TFT screens) is usually a good deal.

2.9 Choosing the Operating system

For basic use, the most common operating system (OS) is the best choice and usually the OS is already included on the computer. Only if you have some particular needs, such as software programming, might you want to change OS.

Apple Macintosh computers have a proprietary operating system (MacOS), now at version 10.1 (OSX); PCs can have any of the Microsoft proprietary operating systems (Windows 98, Windows ME, Windows 2000, and the lastest, Windows XP).

All computers can also support the Linux operating system: Linux is a free Unix-like operating system originally created by Linus Torvalds with the assistance of developers around the world. The source code for Linux is freely available to everyone. Often Linux is used in bio-informatics as a platform to create new software, tools, and web databases.

When choosing a personal computer, a Macintosh or an assembled PC represent the most common choices. Just a few tips to help in the final decision.

Why a Mac?

Pros: It is sold as 'one-piece', its components are chosen for one reason, they represent the best quality/price ratio and can fit together well. The MacOS is very user-friendly, any non-specialist can immediately become an expert user without great effort. The MacOS is very well-organized and self-consistent; OSX also offers Unix-like ability and stability.

Cons: It is more expensive than a PC. It is less common than PCs (10% - 20% Mac market share versus 80% - 90% PCs): this can mean that commercial software might be either not available or more expensive since it is sold as a complete unit, sometimes you get 'extras' that you don't need (e.g. a DVD drive is not used very much in molecular biology labs).

Why a PC?

Pros: A basic PC is less expensive than an 'out of the box' Macintosh (each component can be chosen separately and assembled according to the user needs). Since it is very popular, it is very easy to get help and advice. A lot of software (mainly, but not only games) is available only for Windows systems.

Cons: The OS is usually Windows, which is less user-friendly than it appears. Sometimes it is very difficult to perform even simple tasks without the help of an expert user since it is assembled with pieces which are not built to fit, the overall computer performance may suffer. Windows OS are often less stable than Mac OS.

3

Personal Internet Security

Michele Quondam

Contents

Abstract

Some simple rules and information to avoid the most common problems about viruses, hackers, email attacks, and some general security issues.

From: *The Internet for Cell and Molecular Biologists: Current Applications and Future Potential*
ISBN 1-898486-32-8 © 2002 Horizon Scientific Press, Wymondham, UK

Any computer connected to the internet is vulnerable to **virus** and **hacker** attacks. The only way to make your computer completely secure is to turn it off or disconnect it from the Internet. These solutions present however some obvious inconveniences. An unprotected computer connected to the Internet by cable (University connection or home connection by ADSL) is easier to hack because it is 'always-on' and often has a static (permanent) IP address. This means that once a hacker finds your computer, it is very easy to find it again. Conversely, most dial-up Internet connections (through modem and phone line) use a new IP address each time you connect which makes it much harder to find your computer again.

3.1 What is a Virus?

A computer virus is a piece of software that has been written to enter your computer system and infect your files. A virus can have varying damage properties: from destroying all computer data to simply slowing down the computer. Typically a computer virus will replicate itself and try to infect as many files and systems as possible. If your system is infected, like a human being infected by a virus it can infect other computer systems across networks or floppy disks and email. Viruses are typically of two types: program virus and macro virus. The program viruses are attached to executable files (like .exe files in PC systems) and when the infected program is executed the virus is loaded in the system memory. From the system memory, the virus can infect any other executable file. Macro viruses infect files run by applications that use macro languages, like Microsoft Word or Excel. The virus looks like a macro in the file, and when the file is opened, the virus can execute any command.

Virus penetration perfectly reflects the computer and Operating System distribution on the market: about 85% of viruses attack the Microsoft Windows Operating Systems on PC computer while only 15% are made for Macintosh and Unix Systems. Obviously the macro viruses can infect any computer type where we run applications that use macro languages.

3.2 What is a Hacker?

Universally the word hacker means someone who breaks into a computer system, often on a network, bypasses passwords or licenses in computer programs, or in other ways intentionally breaks computer security. A hacker can be doing this for profit, maliciously, for some altruistic purpose or cause, or only for the challenge of it. In reality this is the definition of a **cracker**. In the computer science environment, the word hacker defines a proficient programmer or engineer with sufficient technical knowledge to understand the weak points in a security system, but who is generally not involved in cracking activities. In common language however, *hacker* is often used instead of cracker.

3.3 Protection Software

Using a personal **firewall** and **anti-virus** software will protect your computer and data from most, if not all, virus and hacker attacks, if you install them correctly and keep them updated. It should however be kept in mind that 'security' is an intrinsically relative concept. You can always increase your security levels, which will make attacks more difficult. However, complete security is never a reality: as an extreme case, for instance, a computer which is disconnected from the net and switched off could be stolen or could burn in a fire. Without going to these extreme situations, you should just be aware that if a computer is connected to a network, a way in always exists. The issue is how much time and skill will be required to find it. Good security means placing a high threshold for that.

3.3.1 Firewalls

A firewall is a software or hardware device that sits between your computer and your internet connection. It watches the traffic that goes back and forth and usually restricts input traffic or warns you about strange or unauthorized traffic from your computer to the internet. For example, many hackers will scan the net to see which network software computers are running before an attack. On the other hand, personal firewalls are often overly sensitive: sometimes even visiting certain web pages can make the firewall think your computer is scanning someone. Installation of a firewall is not necessarily easy and can require professional intervention. Some personal firewalls are easy to install and configure can be downloaded from the net. A good freeware (free software) product called 'ZoneAlarm™', suitable for Windows™ operating systems, is offered for instance by Zone Labs, http://www.zonelabs.com/, which also provides a more complete and customisable commercial solution called 'ZoneAlarm Pro™'.

Another excellent commercial product is the 'Norton Desktop Firewall', available from Symantec (http://www.symantec.com) for both Windows and Macintosh operating systems.

3.3.2 Anti-virus software

Anti-virus software is a program that scans your computer hard drive to find and eradicate viruses. It also scans floppy disks, memory, internet downloads, and email attachments. If a virus is found, the anti-virus attempts to destroy it, or if this is impossible, it quarantines the infected file in a special directory.

It is very simple to find trial versions of anti-virus software over the internet. The best solutions are the Norton package (http://www.symantec.com) or the McAfee anti-virus (http://www.mcafee-at-home.com). After 30 days of evaluation period, it is necessary to purchase the software in order to continue to use it.

It is also possible to scan a computer for viruses directly from the internet without installing a complete anti-virus package; this solution is fast and free, but will not protect the computer from subsequent attacks as a permanently installed anti-virus does; the best online solution is by Housecall at http://housecall.antivirus.com. The advantage of the 'online scanning' solution is that the virus definitions are always up to date.

As a general consideration, given the continuous production and distribution of new viruses, it is essential to keep your anti-virus software updated. Often the updates are in a file that contains the most recent virus definitions that allow the anti-virus software to recognize the latest viruses. Updating your virus definitions file every month or so is a very good idea. Many common anti-virus software packages can be configured so as to update automatically whenever a new virus definition file is available from the web site of the software producer.

If you are connecting two or more computers to the Internet with a single connection, or if you are part of university network, you should also use a hardware router with firewall features (ask your system administrator for details). These products provide adequate protection because most 'attacks' are 'impersonal'. In other words, the attackers often do not specifically target your computer, but simply scan the net in search of an accessible machine, especially if it is part of a university or department network. With firewall protection you can block this scan by making invisible all computers behind the firewall router.

3.4 Email special attacks

E-mail messages can contain or constitute security threats unrelated to viruses. The most common types are bomb and spamming emails.

3.4.1 Bombing

An email 'bomb' is a multi-copy email sent to a particular email address; the large numbers of email slow down the email receiving server and usually make recipient email addresses unusable. Some email bombs contain up to 10,000 - 50,000 identical emails. Blocking bombing can be very difficult and should be done in conjunction with your system administrator or your Internet Service Provider (ISP).

3.4.2 Spamming

Email 'spamming' is a variant of bombing; in this case the same email is sent to hundreds or thousands of users. It's an attempt (often successful) to deliver a message over the Internet to someone who would not otherwise choose to receive it. Almost all spam is commercial advertising, but it may also occur innocently, as a result of sending a message to mailing lists and not realizing that the list explodes to thousands of users, or as a result of an incorrectly set-up responder message.

Potential target lists are created by scanning newsgroups, stealing Internet mailing lists, or simply searching the Web for addresses. This is often done by automated software that recognises and extracts e-mail addresses from web pages.

A number of steps can be taken to fight against or minimize spamming. One of the more effective general strategies is to register for a free e-mail account that will be used for most internet activities and can be just dropped in case of excessive spamming, while using your main e-mail address just for work and for your closest friends. The other very general rule is: never reply to a spammer. This will just tell him that a functional e-mail address was reached and will result in a net increase of junk messages in your mailbox. The same holds true for "removal instructions" often found at the end of spam mails: most likely the address that you are supposed to write to in order to be removed do not exists or will serve as collector of working e-mail addresses.

Here's a short compendium of anti-spam measures, taken (with some modifications) from http://vizier.u-strasbg.fr/~heck/spams.htm, a good starting point to search for info about protection from spamming on the net:

- Never, ever, respond to a spam (this would validate your e-address); rather file a complaint with the spammer's ISP. Note that the ISP is not necessarily easy to identify, as the headers of spam messages will often be manipulated.
- Avoid registering your e-address in any public list and make sure your postmaster is not registering it for you.
- Never request information via e-mail or the web if you are not sure your e-address will not be included subsequently in an e-mailing list.
- Avoid registering your e-address in lists of attendees at conferences, meetings, and so on.
- Never drop a business card with an e-address in an exhibitor collecting box.

- Avoid returning registration/guarantee cards for any kind of product with your e-address.
- Never circulate yourself a list of e-addresses without the explicit consent of the persons concerned.
- Set up filters on your mailing system (but be aware that headers of incoming messages will often be forged).

Other good sources of anti-spam info can be found at:

http://spam.abuse.net/spam/
http://www.spamcop.net/
http://www.ecofuture.org/jmemail.html
http://home.att.net/~penn/spammail.htm

3.5 Simple general security rules

If you follow some simple rules, you can avoid most of the common security problems:

1. Use maximal caution with email attachments, especially with unexpected emails; never open an attachment without checking it with anti-virus software.
2. When you download a file from an unknown web site, again you must scan it.
3. When you don't need to use it, turn off your computer or disconnect it from the network.
4. Open a free email account (like at hotmail.com or yahoo.com) and use it to register on web sites, so your primary email account will be used only by your colleagues and friends. Avoid the posting of your e-mail address to web sites, unless strictly necessary.
5. Do not use your credit card number to verify your adult age, and in any case don't send your credit card number if the web site you are visiting is not trusted.
6. If your computer is used by other people, don't activate the Internet Explorer or Netscape password and data management tools

7.   If you need to send confidential information by email, use software to encrypt the message. The most popular (and free) package is PGP: you can download it at http://www.pgpi.com

4

Design and Build Your Own Lab or Departmental Home Page

Andrea Cabibbo

Contents

Abstract

It is increasingly likely that people wishing to contact you or to have information on your research activities will look for your departmental or personal web page. If have not already, the moment has come to build one. You will see that this is much easier than you might think.

From: *The Internet for Cell and Molecular Biologists: Current Applications and Future Potential*
ISBN 1-898486-32-8 © 2002 Horizon Scientific Press, Wymondham, UK

This chapter is about building web sites. It will be assumed that the reader is not familiar with concepts such as html, web server and FTP; everything will be explained from scratch. After a global overview of the process, enough details will be given on how to plan and build the site to allow the reader to perform all the required steps by himself.

The world wide web was originally based on the Hyper Text Markup Language or HTML, which allows the display of both text and images on a page and provides tools to format the appearance of these elements. At this time, the web was basically a collection of static pages, often containing hyperlinks to other pages, so as to form a real "web" network.

Since these early days, the panorama has been enriched by the appearance of a number of more sophisticated programming tools, such as javascript, java, perl, php, XML and others, that allow a much tighter control of the appearance, function and behavior of web sites, often turning them into sophisticated online applications that allow, for example, searching of complex databases directly over the web and formatting the results according to your needs. This is the case for instance with web sites such as Pubmed, that allow access to Medline and sequence/structure databases.

The following chapter will focus exclusively on building sites the simple, old way, that is by using HTML. The basic concepts can be easily learnt with minimal initial effort. Once the basics are acquired, the reader will be ready to move to more sophisticated implementations.

It should be noted that HTML, despite being simple and old, is extremely powerful and will allow you to publish on the web nearly everything you could think of: text, data, images, downloadable files (documents, multimedia, powerpoint files etc.).

4.1 A global view

As for the construction of all web sites, this is a two-step process:

a) Planning the web site.
No programming or computer skills are required for that. You should decide the overall structure of your site and the kind of information you want to supply to your visitors. It is a good idea to have this very clear before starting to build the web site.

b) Building the site and uploading the site to a web server.
Here you produce the various html pages and graphical elements that compose your site and upload everything to a web server. Now, what is a web server?

The web server: what is it and how do I transfer my files there

A web server is a computer that holds all the files related to your site (html pages, images, downloadable files, multimedia. . .) and sends this material to the visitors of the site when they access the corresponding URL (Uniform Resource Locator, e.g. http://www.yale.edu/yourname/index.html, see Chapter 1) through their web browser. As opposed to the web server, the browser software on the computer of the visitor is called a web client, or simply the client. The web server that will host your lab page will typically be a machine belonging to your institution. In most cases the system manager of your institution will be able to open for you an account on this web server. This means that you will be given access to a personal folder (also called "directory") on this machine. Access to this folder/directory will be restricted by a personal username and password that will be supplied by the system manager, in order to prevent other people modifying or altering your personal files. These data should be kept confidential, in order to avoid the possibility that a visitor to your site might see one day, instead of your site, a nice page saying "hackerguy was here" or worse.

Hundreds of sites exist that offer free web hosting. If you have problems in obtaining a web account from your istitution, you can easily turn to a free hosting such as the one offered by Tripod at http://www.tripod.com) or to the web space that often comes toghether with the internet connection when you sign-up with an ISP.

You will transfer your HTML files and images from your computer to your folder in the web server machine by the so-called "file transfer protocol" or FTP. FTP is implemented by easy to use applications called FTP clients, available for all operating systems. To access your remote folder, you will have to know its address on the web, which will look like ftp.yale.edu or web.nyu.edu/yourname. You insert the address of the folder, together with your username and password, in the appropriate window of the FTP application, press a button, wait a couple of seconds, and here's your folder with its contents displayed to you. In some FTP applications, such as "Cute-FTP" for Windows systems, you will be presented with a bipartite window, in which the left part displays the contents of your hard disk, while the right part displays your folder on the web server. You will select files from the left part (your computer) and move them to the right part (your remote folder) with a click. In other FTP clients, such as "Fetch" for Macintosh systems, a single window appears that shows the remote folder. If you press the "put file" button, a window will appear representing your hard drive, from which you will be able to select the files to transfer. The interfaces of FTP applications are usually very intuitive and easy to use. So, transferring files to the web server is usually an operation that will take you just a few minutes, depending on the number and sizes of your files and on the speed of your internet connection. Many different FTP clients for all operating systems can be downloaded from http://www.tucows.com, for example.

URLs

Your remote folder will correspond to a web address, something like http://www.yale.edu/yourname/, where yourname is the name of your personal folder. The address of the file "index.html" contained in your folder will be: http://www.yale.edu/yourname/index.html. Your folder might contain other folders, called subdirectories. If your folder contains a subdirectory called research, the address of this subdirectory will be http://www.yale.edu/yourname/research/. If the "research" subdirectory contains in turn a file called "results.html" the address of this file on the internet will be http://www.yale.edu/yourname/research/results.html. There is no limit to the number of subdirectories that you can create within your main directory. An effort should be made to organize directories and subdirectories according to rational criteria. For example, all the graphical elements of your site could be contained within a subdirectory called "images", while personal web pages of your lab members could be contained in a subdirectory called "people". The overall organization has no specific rules and very much depends on personal choices. In the simplest scenario, everything can be located inside a single common directory.

What is your web site in practical terms: html pages and graphical elements

Strictly speaking, your web site will be a collection of HTML files and images, contained within an organized tree of folders and subfolders (directories and subdirectories) on the web server. The HTML files are plain text files, in which the text that will appear in your site is contained. This text is usually surrounded by "tags" that determine its format. For instance in the sentence

I was <B>very</B> late yesterday evening and <U>she</U> was already gone,

"very" will be displayed in bold and "she" will be underlined. Actually, the job of web browsers is to understand the tags and format and display pages accordingly. If an image is to be displayed together with the text, a special tag (the <img> tag) will be used for that:

I was <B>very</B> late yesterday evening and <U>she</U> was already gone. <img src="/images/judie.jpg">

This will insert the image file "judie.jpg", contained in the subfolder "images", after the text.

You don't even need to know HTML to produce web pages: very easy to use visual editors exist that resemble graphical applications such as Corel Draw or Microsoft PowerPoint. You will insert text and images, format everything and the application will create the HTML code for you. This topic will be discussed in more detail in the next part of the chapter.

A note on images: images can come in many different formats, such as tiff, pict, jpeg, gif, png, tga and many others. The important thing to know is that basically only three formats are suitable for usage in web pages, namely jpeg, png and gif. If you have images in other formats, you will have to convert them into one of these. We will not go into the details of the different formats here, as this is beyond the scope of this book. Suffice it to say that small, simple images and computer-generated images are often used in png or gif format, while larger, more complex images, such as photographs, are best rendered as jpegs.

Let us now move to the practical part: how do I build my web site?

4.2 Building the web site

4.2.1. Planning the site with pencil and paper

At this stage, just forget about HTML and programming. Attention should be focused on the information that you wish to include in the site and on how this information will be organized in order to be easily accessible to your visitors. It is important to keep in mind that the web has its own "rules". Therefore, your site should **not** be simply an internet transposition of what could be your annual report. This would result in a highly boring site. On the contrary, you should take advantage of the peculiar and unique possibilities offered by the web. Examples of information that you might want to offer include, but are not limited to, the following categories:

- Contact information (address, e-mail, phone, fax. . .)
- How to reach your institute/lab (maps, transport information. . .)

- Members of the department/lab, with summaries of activities/duties
- Ongoing research projects and future developments
- Publication records
- Positions available
- Funding sources
- Downloadable files (lessons and seminars in PDF or PowerPoint format, text files…)

General organization of the site and common mistakes

It is reasonable to dedicate a separate page to each of the categories listed above. Some of the categories could in turn be split into different pages (for example one for each research project or one for each member of the group or for each lab in the department). You will therefore end up with a multi-page site. This brings up the concept of "**navigability**" of a site.

To make this clear, imagine having a main ("home") page that lists the contents of your site and includes links to "contact info" and "ongoing research projects". A visitor, let's call him David, follows the "ongoing projects" link, and then, though a second click, ends up in the "Role of the YourProt-2 gene product in cell-cycle progression" page. Stimulated by the interesting contents, David will want to tell you about a great observation that springs to his mind, likely to turn your project into a "Science" paper with just one key experiment. If the site is poorly planned, in order to do that, David may have to hit the "Back" button of the browser twice to return to the home page, and then another click to reach your "contact info" page. However time is limited, pages take time to load and people are busy. David receives a phone call on extension 2. Your paper ends up in "Cell Cycle Weekly".

In an even worse scenario, you placed the contact info in a page linked to as "about us", "general information" or "other info". The fact that your e-mail address is there is absolutely clear. To you only, though. To find it, David will have to go though an unpleasant trial and error process and, again, time is limited.

This little story exemplifies two important general concepts:

- All the information in your site should be available through preferably a single click, maximum 2 clicks (this is a widely recognized rule) from every location of your site. A simple way to achieve this will be to include in every page of your site a navigation bar, typically at the top or on the left side of the pages, linking to all the other pages or main sections of your site.

- The names of the links of the sections should clearly and directly reflect the contents.

Once you have planned in detail the structure of your site (sections, pages, internal navigation tools), you can move to the next step, that is building the site itself.

4.2.2. Building the site

You can design your site without having to write a single line of HTML. A number of WYSIWYG (what you see is what you get) visual editors exist that will allow you to compose a page with simple tools similar to the ones present in common graphical applications and will convert the page to HTML. These will be discussed in the next section. However, HTML is so incredibly simple that it is definitely worthwhile spending the five or ten minutes required to learn the basics. This can often be useful to overcome by hand some limitations of the visual editors. The HTML basics will be covered in the last section of this chapter.

4.2.2.1 Visual HTML editors

Many visual HTML editors are available on the market. They range from very simple editors, such as Netscape Composer (http://home.netscape.com/communicator/composer/), and Microsoft "FrontPage" (http://www.microsoft.com/frontpage/), which basically can only handle HTML (FrontPage is somewhat more sophisticated, visit the web site for details), to more advanced products such as Adobe GoLive (http://www.adobe.com/products/golive/) or Macromedia Dreamweaver (http://www.macromedia.com/software/dreamweaver), that have a wide range of additional features such as the possibility to automatically add java applets, javascript code or to build database-driven web sites. These high-end products are however significantly expensive and might be "oversized" and complex for a beginner.

We will focus here on the usage of Netscape composer as it is completely free, very simple to use and is embedded in the "Netscape Communicator" suite, easily downloadable from the Netscape web site. If you have Netscape Communicator installed on your machine, you already have composer. Note that the interface of other "low range" HTML editors is often conceptually very similar to composer.

Before you start, **create a new folder** somewhere in your hard drive. You will **use this folder to store all your HTML files and images** related to this short tutorial.

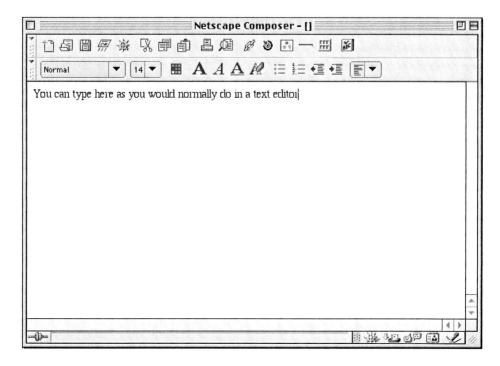

Figure 4.1. A composer window.

To start composer, launch Netscape Communicator and select, from the "edit" menu, new → blank page. A composer window will open, as shown in Figure 4.1 in which you can type as you would normally do in a text editor. Type something in the page and save the file as index.html **in the folder created previously**. It is important to include the extension ".html" or ".htm" in the name of the file. That's it, you have created your first html document! You can view your file in either Netscape or Explorer, by opening the file from the "edit" menu. To preview the file in Netscape, you can simply press the preview button on the top of the composer window (the fifth from the

Figure 4.2. A pop-up window in which you will be able to enter the URL.

left in the top row of buttons). While you progress in the construction of your site, you will find yourself in a cycle in which you modify a page, save and preview, modify, save and preview. It is important to preview as, despite composer being a WYSIWYG editor, there are always significant differences in what you see in the Composer window and how the file is displayed in the browser window. As a general rule, it is an excellent idea to preview your site, at least at the end of the project, with different browsers and possibly in different operating systems. You might be surprised, for instance, by the huge difference between viewing the very same site with Netscape on a Macintosh and with Explorer on a PC. For instance, in a PC, text will look approximately 1/3 bigger that in a Mac. Worth a try.

How to add an external link

In your index.html file, add the words "Yale University", then select these words and press the button with a little chain on the top of the composer window (See Figure 4.2). A window will pop-up in which you will be able to insert Yale's URL, as shown in Figure 4.2. Press the OK button. "Yale University" is now a link to Yale and will show as blue underlined text. You can save, preview in Netscape and follow (click) the link: it should bring you to the Yale web site.

Creation of a link to another page of your site

Let us now create a new page in our folder, named page2.html. To do that, go to edit→new blank page, write something in the page, save in the usual folder as page2.html and close. How can we create a link to page2.html from the index.html page? In the page index.html, write "go to page 2", select, press the chain button and then, instead of writing down an URL, press the "choose file" button. This will open a window from which you will be able to navigate your hard disk and select the newly created page2.html file. Press OK, save the changes and preview in Netscape. If you now follow the "go to page 2" link in index.html, you will navigate to page2.html.

This is basically all you need to know about links. With variations on these very simple operations, you can easily create a full featured multi-page web site. An example of such a basic design is the http://cellbiol.com site. To better understand the organization of this site, you might access the following mirror of cellbiol.com: http://homepage.mac.com/cellbiol/index.html. All files composing the site are contained within the "cellbiol" folder of the web server that hosts http://homepage.mac.com. Another example of such a basic html site can be viewed at http://w3dibit.hsr.it/molecular_immunology/. In this case some javascript was used to create a few "rollover" effects (an image that changes when you pass over it or over a link with the mouse), but the structure of the site is very simple, with a total of 5 html pages (home, people, projects, publications, links) all contained within a single folder called "molecular_immunology" and a number of images, all contained in a subdirectory of molecular_immunology called "images".

How to format text

Several text formatting options are available in html. In composer, some of these are located in the lower part of the top of the composer

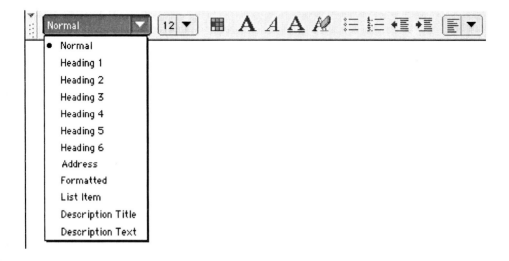

Figure 4.3. Text formatting options.

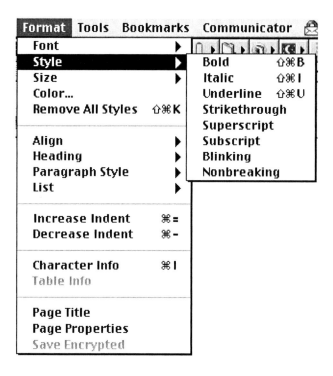

Figure 4.4. The "Format" menu: A more complete set of options.

window (Figure 4.3). A more complete set is located in the "Format" menu (Figure 4.4). Text can be bold, italicized, underlined (not recommended, should be reserved for links), superscripted, subscripted. Text can be colored.

Text font can be selected. However, it is important to realize that in order to be displayed, a font must be present on the client computer, the one of the visitor to your web site. Therefore there's no point in selecting rare, exotic fonts that you might have installed in your computer as these will be substituted by other fonts in the client computer. As a result, your pages will look very different from your expectations. For this reason, you might want to limit the selection of fonts to the most common ones, such as Times, Georgia, Geneva, Courier, Arial, Helvetica and a few others.

The "heading" tag is often used to create a hierarchy of font sizes, heading 1 being the bigger size and heading 6 the smallest (see Figure 4.7). This is an incorrect use of "heading" however and may lead to unexpected results. The use of "heading" should be restricted to the logical division of content. For example, it is appropriate to format the title at the head of a web page as heading 1. Major divisions of the text (like chapters in a book or sections of a paper) can then be heading 2, and subdivisions of those heading 3. It is illogical to use (for example) a level 4 heading for a chapter title and a level 2 heading for a section within that chapter, no matter what it looks like on your particular set-up. Preserving heading consistency will ensure that the web page behaves as intended on different operating systems and browsers, independently of the default font set in the client.

A bullet list or a numbered list of items can easily be created by selecting text and pressing the relative button on the top bar (Figure 4.7).

There are two other basic things that you should know in order to build a nice-looking web site: how to insert images and how to insert and format tables. Here we go.

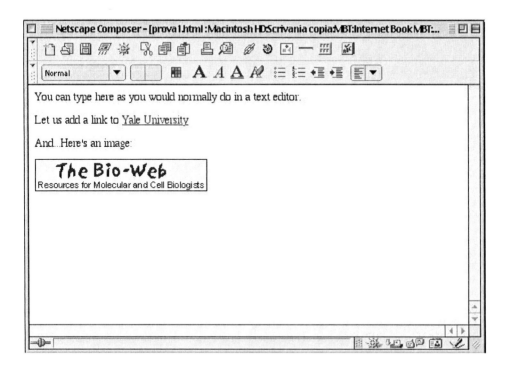

Figure 4.5. Adding a graphic to your web page.

Figure 4.6. The "insert image" window of composer.

How to insert images

Put an image file **in the usual folder**. If you do not have any, just go to a random web site, click on an image and drag the image on your desktop or "right click" on the image (on a Macintosh, place the pointer on the image and keep the mouse button pressed until a menu appears) and select "save this image as". This will copy the image on your disk. Then move the image to your folder. To insert this image in the index.html page, open the page in composer, position the pointer where you want to insert the image and click the button with a little triangle, square and rhomb on the top of the composer window. A window will pop-up with a button "browse files". Press this and navigate your hard disk until you can select your image, then press ok and save. In this example, a Bio-Web banner was added to index.html (see Figure 4.5).

You have several options to position the image with respect to the page and to the surrounding text and can select from them in the "insert image" window of composer. Figure 4.6 shows a detail of the window, that renders well the concept.

The best thing to learn how the different options work is to play around with them. Composer is unable to render image positioning: you should always preview the file in a browser to check what is going on.

How to insert and format tables

Tables are one of the most potent formatting tools available in HTML. The table editor is accessible from the icon representing a little table in the top bar in composer (where the mouse pointer is located in Figure 4.7). Upon clicking the button, you will be presented with a window in which you will be able to select a number of parameters

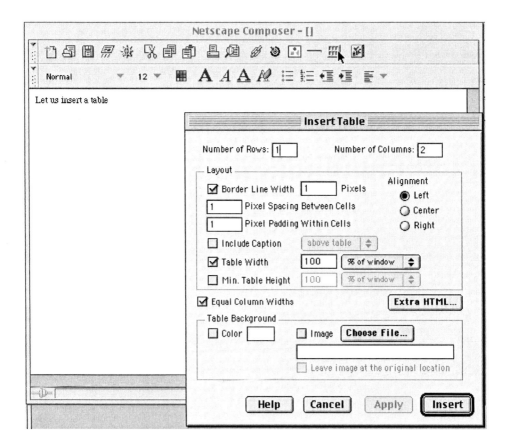

Figure 4.7. The table editor is accessible from the icon representing a little table in the top bar in composer.

regarding the table that you are creating (Figure 4.7).

The main parameters are:

- Number of rows and columns
- Presence of a table border and "thickness" of this border measured in pixels. If you want the border to be invisible, you should put 0 as border line width or uncheck the relative radio button
- The alignment of the table in the page: left, centre or right and the width of the table. You can choose to express this value in absolute (pixels) or relative (percentage) terms. The latter is preferable as the table will more easily adapt to different screen resolutions and window sizes.
- Spacing: the spacing between cells, measured in pixels
- Padding: the spacing between the text within a cell and the borders of the cell, measured in pixels

An example of how to use a table to format and align text and figures can be found at http://w3dibit.hsr.it/molecular_immunology/projects.htm. In this case, a table was used to create, on the left part of the web page, a section in which a small index and a number of figures are located. Each figure refers to a text paragraph located in the right column of the table and each figure and corresponding text are located in a common row. This allows the correct alignment of each figure with the corresponding text, which would be very difficult to achieve without the use of a table. It should be mentioned that the advent of Cascading Style Sheets (CSS) has provided a more powerful and flexible page formatting alternative with respect to tables.

How to make files downloadable from your website

In order to make files downloadable from your web site, the files should be uploaded to the web server. You can them link to them as you would link to another html page. If the files have particular extensions like .doc, .rtf, .hqx, .zip, .bin and others, upon clicking the link, the browser will automatically offer to download the file instead of trying to display it in a browser window. A note on this topic is also present in chapter 12.

Editing the source code of your HTML files generated with Composer

From Composer, you can edit the HTML by hand. This can be useful to manually adjust the code and to add "exogenous" HTML code, for instance to insert a form or a counter in your pages (see next paragraph). From the edit menu, select "html source". The first time you do that, composer will ask you to select a text editor for editing the source. Select "Simple Text" on Macintosh or "Notepad" on Windows systems. The HTML file will then be opened and viewed as a plain text file. You will be able to edit this code manually. As an alternative, you can directly open HTML files generated with Composer (of with other editors) from within any text editing application. To simply view the HTML source in composer, without editing, go to the "view" menu and select "page source". HTML is discussed in more detail in section 4.2.2.4.

4.2.2.3 Bells and whistles

A web site can easily be enriched with a number of advanced features, such as for example:
- Web forms to be filled in by the visitors to the site. These can have a number of different purposes, such as for example obtaining feedback on the site, requesting information and others. In the simpler implementation of a web form, when the user presses the OK or SEND button, a message with the contents of the form is sent to an e-mail address specified by the creator of the site
- A counter that keeps track of the number and source (IP address of the visitor) of hits to pages of your site. This can be very interesting, as it allows you to keep track of where your visitors are coming from (country, institution. . .)
- A site-specific search engine that allows visitors to search for

keywords within your web site only
- A bulletin board in which visitors to your site can discuss about selected topics by posting messages
- Many others. . .

The good news is that you can implement all of these services very easily, without having to write a single line of new code. The reason is that a number of web sites provide free tools for webmasters that allow to create all these services in your site by providing few lines of code that are to be copy-pasted in your HTML and will do all the job. We will not go into the details of how every single service is implemented, but will rather list a few of these service providers and the main tools they offer. Most of these sites are designed to be used by non-professional webmasters and are therefore rich in explanations and tutorials on how to use the services they provide. Most often, in exchange for the tools provided, these sites ask in exchange to display a link or a banner that directs to their site. Elimination of these banners can sometimes be obtained for a small annual fee.

Bravenet: http://www.bravenet.com.
Offers basically all the services (more than 30) you might think of: guestbooks, forms, counters, forums, search engines, mailing lists and many others. The quality of the tools is somewhat variable, but a good degree of customization is nearly always available. The web form for the submission of links to http://cellbiol.com is an example of customization of a bravenet form (http://cellbiol.com/add.htm).

Atomz: http://www.atomz.com.
Offers one of the best site-specific search engines available on the web, for free. The search results page is entirely customizable. You can easily render this page similar to the rest of your site, giving visitors the feeling that they are not leaving your site while performing a search.

Digits.com http://www.digits.com and **The Counter** http://www.thecounter.com.
Both sites provide a free counter to keep track of the number of visitors to your site pages. For example, the counter displayed in bottom of the home page of http://cellbiol.com is from Digits.

Hits4me.com http://www.hits4me.com.
Password system to restrict access to portions of your web site, web polls, guestbooks and more.

Many other such sites exist. A Google (http://www.google.com) search with the string "free webmaster tools" will provide an exhaustive list.

4.2.2.4 Short course of HTML: the basics

The basics of HTML can readily be acquired in 10 minutes, without previous knowledge of any programming language whatsoever. In a way, more than as a programming language, HTML can be viewed as a formatting tool. Instead of going through theoretical considerations, we will proceed through a few easy examples:

Let us consider this code:

Example 1: a simple HTML document

```
<HTML>

<HEAD>
<TITLE>The Cell-Cycle Lab Page</TITLE>
</HEAD>

<BODY>
<P>
Welcome to the Cell-Cycle Lab Page. This page is still very basic but
we are glad that you visited
</P>
</BODY>

</HTML>
```

Type down this code, exactly as it is, with a basic word processor such as SimpleText if you use a Mac or Notepad on a PC with Windows.

Save the document as "example.html" in a dedicated folder. If you use a more sophisticated application such as Microsoft Word (not recommended), be sure to save the document in "Text Only" format. Then launch your favourite browser and open the example.html document from within the browser application to observe the result. You will then be able to go through a cycle in which you modify the code with the text editor, save the changes (very important!), and reload/refresh the page in your browser window to see the changes.

Note the following:

• A text within brackets (<TAGNAME>) is called a tag. Most tags come in pairs: a start tag (<HEAD>, or <P>) and an end tag (</HEAD> or </P>)

• Everything in the page is included between the <HTML> and </HTML> tags. This tells the browser that the document is an HTML document.

• HTML documents are always divided in a "HEAD" and a "BODY" section

• The HEAD section can contain different elements, but always contains the title of the document. As shown in example 1, the title is enclosed within a pair of TITLE tags (<TITLE>The Title goes here!</TITLE>), and will show at the top of the browser window.

• The actual contents of the page will be located in the BODY section.

These notions are indeed all you need to know in order to write down a simple HTML web page.

Let's now add a link to the Massachusetts Institute of Technology in our page:

Example 2: linking to external and internal pages

```
<HTML>

<HEAD>
<TITLE>The Cell-Cycle Lab Page</TITLE>
</HEAD>

<BODY>
<P>
Welcome to the Cell-Cycle Lab Page. This page is still very basic but
we are glad that you visited
</P>
<P>
Our Favorite Institute: <A HREF="http://www.mit.edu">The MIT</A>
</P>
</BODY>

</HTML>
```

To add a link, the <A> tag is used. The general formulation is:
<A HREF="The URL of site here">words to be linked here</A>

To add a second link on the same line of text:
Our Favorite Institute: <A HREF="http://www.mit.edu">The MIT</A>
But we also love <A HREF="http://www.yale.edu">Yale University</A>

To make a list of links in a column the <BR> tag can be used. The

 tag is always single and does not need to be "closed" by a
terminator tag:

<A HREF="http://www.mit.edu">The MIT</A>
<A HREF="http://www.yale.edu">Yale University</A>
<A HREF="http://www.nih.gov">The NIH</A>

A more elegant way of doing this is to use the "unordered list" tag, <UL>, with "list items" <LI>:

```
<UL>
<LI><A HREF="http://www.mit.edu">The MIT</A></LI>
<LI><A HREF="http://www.yale.edu">Yale University</A></LI>
<LI><A HREF="http://www.nih.gov">The NIH</A></LI>
</UL>
```

Linking to internal pages of your site

To link to internal pages the <A> tag is used, as for external links.

- If the page to be linked to is in the same directory as the page that contains the link, the link will look like:
  <A HREF="page2.html">Click here to go to page 2</A>

- If the page to be linked to is in a subdirectory (with respect to the directory where the page that contains the link is located) named "research", the link will be:
  <A HREF="research/page2.html"> Click here to go to page 2</A>
.

  research/page2.html is also known as "the path" to page2.html. The path is relative to the directory of the page that contains the link.

- If the page to be linked to is in a directory that is located up in the directory hierarchy, the ../ expression is used. For instance, if you want to link from a page located in a folder to a page located in the parent folder of this folder, the code will be:

  <A HREF="../page2.html"> Click here to go to page 2</A>

Formatting text

To have a text displayed in bold: <B>text here</B>
Italics: <I>text here</I>
Underlined: <U>text here</U>

To make a bullet list:
<UL>
<LI>item 1</LI>
<LI>item 2</LI>
<LI>item 3</LI>
</UL>

Many of the previously favoured ways of formatting HTML, especially those to do with appearance and page layout, have been deprecated so as to maintain user-agent (client) compatibility. People are no longer using just Netscape, Lynx or Explorer, but WebTV, PDAs, WAP-enabled mobile phones and a host of other browsers to access the WWW and the internet. Thus is it important to ensure that your site is accessible. The best way to do this is to adhere to the recommendations laid down by the World Wide Web Consortium (W3C; http://www.w3.org/) and to validate your code using one of the online tools (see below). As mentioned previously, a powerful and sophisticated way of formatting your pages is to use CSS. CSS is more complicated than HTML, but tutorials are available (http://www.htmlhelp.com/reference/css/).

How to insert images

- External images
  <img src="http://www.asite.com/images/image1.jpg" width="100" height="250">

- Internal images
  <img src="/images/image1.jpg" width="100" height="250">

HTML reference sites

The simple elements of HTML described so far are enough to build a nice looking web page. HTML however offers many possibilities not described here. For more on HTML and on web pages building we suggest the following web sites:

The official HTML pages and recommendations
http://www.w3.org/MarkUp/
http://www.w3.org/TR/#Recommendations

The Web Design Group pages
http://www.htmlhelp.com/
http://www.htmlhelp.com/reference/

A beginners guide to HTML:
http://archive.ncsa.uiuc.edu/General/Internet/WWW/HTMLPrimerAll.html

How to build a web page
http://www.darien.lib.ct.us/webclass/default.htm

Cascading Style Sheets
http://www.htmlhelp.com/reference/css/

Writing HTML: A tutorial for creating web pages, subdivided in lessons (excellent)
http://www.mcli.dist.maricopa.edu/tut/

HTML online reference guide. Tags are listed by name or by html tag
http://www.webspawner.com/cc/html/alpha.htm

Dave Raggett's guide to html
http://www.w3.org/MarkUp/Guide/ and
http://www.w3.org/MarkUp/Guide/Advanced.html (more advanced stuff)

Validators
HTML: http://validator.w3.org/
CSS: http://jigsaw.w3.org/css-validator/

5

Using Search Engines and PubMed Effectively

Andrea Cabibbo

Contents

Abstract

It is estimated that at present more than one billion web pages exist, and thousands of new pages are created every day. In this scenario, finding specific information seems very difficult. However, thousands of 'indexes', 'directories' and 'search engines' exist that attempt to categorize the contents of the internet by various means. The directories range from argument-specific ones, such as for instance biological directories or architecture directories, to global directories that attempt to review all possible contents. A typical example of this latter type is the Yahoo directory

From: *The Internet for Cell and Molecular Biologists: Current Applications and Future Potential*
ISBN 1-898486-32-8 © 2002 Horizon Scientific Press, Wymondham, UK

(http://www.yahoo.com/). In Yahoo, all content is arranged into 14 parent categories (e.g. Art and humanities, Business and economy), each of which is subdivided into subcategories, in turn subdivided into sub-sub-categories, down to very specific subjects. For instance, information about PCR in Yahoo has the following path: Home>Science>Biology>Molecular_Biology>PCR. In search engines, the contents are not pre-distributed in categories but rather are searched by keywords. In this chapter we will provide essential information on directories and search engines, together with tips on how to use these resources efficiently, in order to find the right needle in the internet haystack. We will also briefly review the PubMed Boolean search syntax that allows very precise searches for specific research articles.

5.1 Directories and search engines

A clear distinction is to be made between directories and search engines.

5.1.1 Directories

Directories are man-made and sites are added to directories by hand. Each directory has a team of editors that review submitted web sites, decide whether a particular site should be included or not and eventually add the site to the relevant category within the directory. A few years ago, a project called the 'Open Directory Project' (DMOZ) started, with the spirit of the 'open source' projects. This is a non-profit project and the editors are volunteers from all around the world, who collaborate over the net to generate what is potentially the largest human built directory of the web. The original project can be accessed at http://www.dmoz.org/. The Open Directory Project is free to use for everybody and you can even freely copy the contents of the directory and offer them to the visitors of your web site. Indeed, many web sites, such as Netscape Search, AOL Search, Google, Lycos, HotBot, DirectHit and hundred of others offer the contents of the DMOZ directory. With respect to Yahoo, DMOZ has the advantage of being advertising-free and commercially unbiased.

5.1.2 Search engines

Search engine contents are built automatically. Each search engine sends around little applications called robots, spiders or crawlers, that visit web pages and follow all the links in these pages to find other pages in an endless search of new pages and content around the web. Visited pages are indexed according to variable criteria and the indexing data are sent back to the search engine, to be included in the possible outcome of searches from these engines. There is a huge variability in the algorithms used to index the pages and to rank the results for a particular search. This means that the very same search might yield quite different results in different search engines. This is also very true for a different reason: despite appearances, if you search for the string 'DNA sequence analysis' on Altavista and on Google, you are not searching for the same thing. Altavista understands that you are searching all the pages that contain 'DNA' or 'sequence' or 'analysis'. Pages that contain all three terms at the same time are ranked higher, but pages that only contain the word 'analysis' are also returned in the results. Instead Google understands that you are looking for pages containing 'DNA' and 'sequence' and 'analysis', and will then most likely return more relevant results. You can also have Altavista search for pages containing 'DNA' and 'sequence' and 'analysis', but you should then use a particular syntax to perform this search. This is covered in the next section of this chapter.

The SearchEngineWatch web site focuses on all aspects of search engines and web searches. It is located at http://searchenginewatch.com/. A list and short description of the major search engines can be found at http://searchenginewatch.com/links/major.html/. An overview of the features, key search commands and search assistance features of the main search engines can be found at http://searchenginewatch.com/facts/ataglance.html/.

Google

A particular note should be made here about Google (http://www.google.com/). Google has a particular algorithm for ranking the results of your search: it will rank more highly those web sites that are more linked to from other web sites. It is a fact that important or very useful sites are often linked to from other sites. For instance, there are more than 16,000 pages linking to the NCBI web site that hosts the very popular PubMed service. So, having many pages that link to a site is clearly an indication of the importance and of the usefulness of this site. Because of this, Google is really special in returning relevant results for a search. We strongly advise you to start from Google for any search that you might have to perform, as it is by far the best search engine available at the moment on the web. Google has also a special feature that allows you to restrict your search to images or to newsgroups messages. You can access those from the Google home page or at http://images.google.com/ and http://groups.google.com/, respectively. At http://directory.google.com/ you will find the Google version of the DMOZ directory (see 5.1.1). As shown in figure 13.5 (chapter 13), Google also allows the automatic translation of web sites.

5.2 Search syntax: the mathematics of search engines

By using a simple syntax, you can specify to most search engines how your keywords should be combined in order to get the most relevant results. This will work with most search engines: if you want pages that contain all and every term of your search, you can use the '+' symbol. For instance, an Altavista search with the string '+DNA +sequence +analysis' will return only pages that contain all the three terms. However, these terms might be present on different sections of the page. A page containing the sentence 'this page is not about DNA' and later, 'the sequence of events was rapid', and then 'analysis of the data was not performed' will score positive. The solution is to present your query between quote marks: if you search for "'DNA sequence analysis'", only pages that contain the three terms, in this order, will score positive, narrowing your results to the most relevant results. The '-' symbol can be used to exclude pages that contain a specific term.

These concepts can be very useful while looking for a person on the internet. If you are looking for David Baltimore on the net, you should not search for 'David Baltimore' but rather for "'David Baltimore'". This will avoid an endless number of pages talking about, for instance, the summertime trips of David Blisset and wife to Baltimore.

Golden Rule:

IF YOU ARE LOOKING FOR DAVID BALTIMORE, SEARCH:

"DAVID BALTIMORE"

rather than:

DAVID BALTIMORE

Using just a little bit of syntax can make the difference between finding what you are looking for or not.

5.3 Searching for scientific literature: the NCBI PubMed site

The NCBI PubMed site (http://www.ncbi.nlm.nih.gov/entrez/) has become a worldwide reference for searching scientific literature and articles. Figure 5.1 shows what the PubMed input box looks.

You can enter terms, press go, and will get results. However, sometimes one is looking for a very specific combination of terms, for instance papers from a specific author in a particular journal or about a particular subject. In order to perform this kind of search

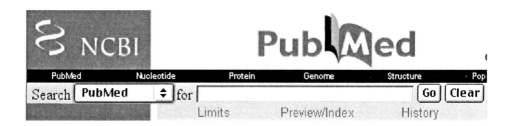

Figure 5.1. The PubMed input box.

efficiently, it is convenient to learn some basic PubMed syntax. Here are some pointers:

- Authors should always be entered as family name followed by initials. For instance Andrea Cabibbo would be 'cabibbo a' and Richard P. Grant would be 'grant rp'.

- You can use Boolean operators (these establish the relative usage of two terms) such as AND or OR, which should always be capitalized in order for PubMed to understand that they are indeed operators. AND among different terms means that all of them must be contained in the search results; the OR operator is to be used to select PubMed entries containing even only one of the listed terms.

- MESH terms are 'Medical Subject Headings'. Medical Subject Headings form a controlled vocabulary of biomedical terms, which is used to describe the subject of each journal article in MEDLINE. The MeSH database contains more than 19,000 terms and is updated annually to reflect changes in medicine and medical terminology. MeSH terms are arranged hierarchically by subject category with more specific terms arranged beneath broader terms.

Example 1

Let us consider a simple example. Imagine you are looking for papers from Andrea Cabibbo in the journal *Blood*. An intuitive possibility would be to enter 'cabibbo a AND blood'. Note that if you simply enter 'cabibbo a blood', PubMed will insert an AND operator by default. This search will not yield the required result as 'blood' will be interpreted as a MESH term rather than a journal name. So, 'cabibbo a AND blood' will actually yield one article, which has 'blood' as an associated MeSH, but no articles from the *Blood* journal. Here comes the syntax: you should specify the category to which your input terms belong, such as author name, journal name, MeSH. This is done by using tags which are placed after the term to be searched. Some of the main tags are:

[au] Author
[la] Language
[dp] Publication date
[pt] Publication type
[mh] Mesh term
[ta] Journal title.

Let us now rephrase our search for publications from Andrea Cabibbo in the *Blood* journal:
'cabibbo a [au] AND blood [ta]'
This will now indeed yield my single *Blood* paper.

Example 2

We now want to list all the review papers by David Baltimore:
'Baltimore d [au] AND review [pt]' will yield all the review papers by this author.
If we want to restrict the search to reviews published in 1998, the search becomes:

'Baltimore d [au] AND review [pt] AND 1998 [dp]'

If we want to specify a time interval between two dates, the dates should be separated by a semicolon and followed by the [dp] tag:

'Baltimore d [au] AND review [pt] AND 1990:2000 [dp]'

Example 3

Parentheses can be used to assign a logical grouping between search terms. We want now to list all the review papers by Neil J. Bulleid or by Stephen High published in the last 10 years:

'(high s [au] OR bulleid nj) AND review [pt] AND 1991:2001 [dp]'

This will yield a very different result from:

'high s [au] OR (bulleid nj AND review [pt] AND 1991:2001 [dp])'

These simple concepts will allow you to perform targeted searches on PubMed. Many other syntactic possibilities and subtleties exist beyond these simple examples. To explore the subject further, we suggest that the reader refers to the excellent PubMed tutorial available at: http://www.nlm.nih.gov/bsd/pubmed_tutorial/m1001.html. Details on Boolean search strategies can be found at: http://www.ncbi.nlm.nih.gov:80/entrez/query/static/help/pmhelp.html.

6

Online Tools for Basic Sequence Manipulation, Restriction Analysis, PCR Primer Generation and Evaluation

Andrea Cabibbo

Contents

Abstract

The analysis of biological sequences often requires some preliminary basic manipulations. For instance it is often necessary to obtain the complementary sequence to a DNA sequence, to reverse a sequence, to get a list of the restriction enzymes cutting sites in a sequence, to translate a DNA sequence to a protein

From: *The Internet for Cell and Molecular Biologists: Current Applications and Future Potential*
ISBN 1-898486-32-8 © 2002 Horizon Scientific Press, Wymondham, UK

sequence, and so on. **Many tools are available online to perform all these operations easily. Often more than one possibility is available to the user. We list here a number of tools freely available online. These and other links are also reported in the "sequence analysis tools" section of the Bio-Web, at http://cellbiol.com/.**

6.1 Restriction analysis

Web Cutter
http://www.firstmarket.com/cutter/cut2.html
Performs restriction mapping of your DNA sequence. A really valuable feature is the ability to detect silent restriction sites, that is sites that can be activated by changing a few nucleotides, without changing the sequence of the encoded protein. Here is a short description of the main features of the program:

- Rainbow cutters: Highlight your favourite enzymes in colour or boldface for easy at-a-glance identification
- Silent cutters: Find sites which may be introduced by silent mutagenesis of your coding sequence
- Sequence uploads: Input sequences directly into Web Cutter from a file on your hard drive without needing to cut-and-paste
- Degenerate sequences: Analyse restriction maps of sequences containing ambiguous nucleotides like N, Y, and R.
- Circular sequences: Choose whether to treat your sequence as linear or circular
- Enzyme info: Click into the wealth of references and ordering information at New England BioLabs' REBASE, directly from your restriction map results
- Automatic sequence search-and-entry from NCBI's GenBank
- Easy, customizable interface and clean and simple results format.

SeqCUTTER from MBS
http://www.mbshortcuts.com/cutter/index.htm
Finds commercial restriction enzyme recognition sequences in your DNA sequence.

Hierarchical identification of silent restriction sites
http://www.science.mcmaster.ca/biochem/hipsrs/public_html/HIP/index1.html

6.2 Basic sequence manipulation

BCM Reverse complement tool:
http://dot.imgen.bcm.tmc.edu:9331/seq-util/Options/revcomp.html
Input a nucleic acid sequence and get the reverse complement as output

EBI Translation Machine:
http://www2.ebi.ac.uk/translate/
Translates DNA-RNA sequences to protein sequences. You can select one of the six frames and the appropriate genetic code.

MBS Translator:
http://www.mbshortcuts.com/translator/index.htm
Translates a DNA sequence. You can select one, three or six frames. Nice output.

BCM Six frame translation tool:
http://dot.imgen.bcm.tmc.edu:9331/seq-util/Options/sixframe.html
Translates the six frames of a sequence at the same time. You can exclude selected frames.

ORF Finder:
http://www.ncbi.nlm.nih.gov/gorf/gorf.html
Finds open reading frames in a sequence. A smart graphical output allows identification of the most convincing coding region by eye. See the example in Figure 6.1.

ORF Finder (Open Reading Frame Finder)

Anonymous

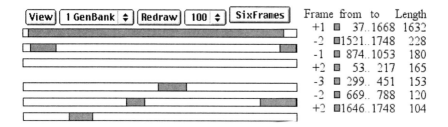

Frame	from	to	Length
+1	37	1668	1632
-2	1521	1748	228
-1	874	1053	180
+2	53	217	165
-3	299	451	153
-2	669	788	120
+2	1646	1748	104

Figure 6.1. The ORF finder at http://www.ncbi.nlm.nih.gov/gorf/gorf.html.

IBCP Color sequence:
http://npsa-pbil.ibcp.fr/cgi-bin/npsa_automat.pl?page=/NPSA/npsa_color.html
Colours residues in your protein sequence based on hydrophobicity, charge, and other features of your choice

CloneIt: http://topaze.jouy.inra.fr/cgi-bin/CloneIt/CloneIt
An online program finding sub-cloning strategies, in-frame deletions and frameshifts using restriction enzymes and DNA polymerases.

6.3 PCR primers generation and analysis

Many modern Molecular Biology techniques rely on the use of oligonucleotides for instance as probes, anti-sense reagents, PCR primers. Here is a list of free online tools to analyse oligonucleotides and to select optimal PCR primers:

Oligonucleotide calculator at NOAA:
http://www.nwfsc.noaa.gov/protocols/oligoTMcalc.html
Calculates melting temperature, %GC, picomoles/OD. See Figure 6.2.

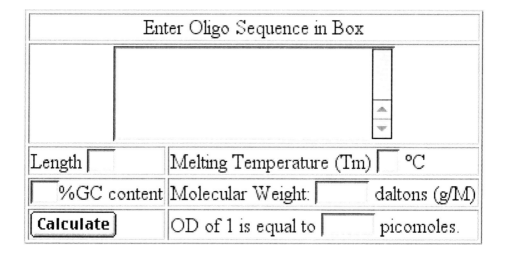

Figure 6.2. Oligonucleotide calculator at NOAA.

Oligonucleotide calculator at MBS:
http://www.mbshortcuts.com/biotools/oligo.htm
Very similar to the previous tool

Primer3:
http://www-genome.wi.mit.edu/cgi-bin/primer/primer3.cgi
Helps in the selection of optimal PCR primers to amplify an imput sequence. We have a personal excellent experience with this tool. Invariably provides working primers. A new interface is available at http://www-genome.wi.mit.edu/cgi-bin/primer/primer3_www.cgi

Primefinder:
http://eatworms.swmed.edu/~tim/primerfinder/
Can be used both to select optimal primers to amplify an imput sequence and to analyse primers already designed by other means for dimer and hairpin formation. This second option is available at:
http://eatworms.swmed.edu/~tim/primerfinder/check.cgi
Figure 6.3 provides an example of hairpin and dimer analysis on a random primer.

Hairpin

```
Oligo, 3 bp (Loop=7), delta G = 0.5 kc/m

 5'  TCAGGCTTGT─┐
      │││  │    T
 3'  AGGTCGTACAG─┘


Oligo, 3 bp (Loop=7), delta G = 0.6 kc/m

          5'  TCAGGC─┐
              │││    T
 3'  AGGTCGTACAGTTGT─┘


Oligo, 2 bp (Loop=4), delta G = 1.4 kc/m

 5'  TCAGGCTTGTT─┐
      ││   ││ ││ │
 3'  AGGTCGTACAG─┘
```

Self-Dimer

```
4 bp, delta G = -5.4 kc/m (bad!) (worst= -41.4)
5' TCAGGCTTGTTGACATGCTGGA 3'
                │││││
        3' AGGTCGTACAGTTGTTCGGACT 5'


3 bp, delta G = -3.6 kc/m (worst= -41.4)
  5' TCAGGCTTGTTGACATGCTGGA 3'
      │││ │        │ │││
 3' AGGTCGTACAGTTGTTCGGACT 5'


3 bp, delta G = -3.5 kc/m (worst= -41.4)
          5' TCAGGCTTGTTGACATGCTGGA 3'
              │││      │││
 3' AGGTCGTACAGTTGTTCGGACT 5'


3 bp, delta G = -3.3 kc/m (worst= -41.4)
5' TCAGGCTTGTTGACATGCTGGA 3'
    ││  ││ │││ │││ ││  ││
 3' AGGTCGTACAGTTGTTCGGACT 5'
```

Figure 6.3. An example of hairpin and dimer analysis on a random primer.

6.4 Sequence analysis servers and links

The following servers offer a wide range of tools for DNA and protein sequence analysis:

Expasy
http://www.expasy.ch/
'The ExPASy (**Ex**pert **P**rotein **A**nalysis **Sy**stem) proteomics server of the Swiss Institute of Bioinformatics (SIB) is dedicated to the analysis of protein sequences and structures as well as 2-D PAGE'. One of the most complete set of tools and databases on the net.

EMBOSS via the PISE interface
http://www2.no.embnet.org/Pise/
Web interface for a wide range of freely available sequence analysis programs.

NCBI tools for bioinformatics research
http://www.ncbi.nlm.nih.gov/Tools/

Amos' WWW links page
http://www.expasy.ch/alinks.html
One of the first and most complete pages on biological resources and databases on the net. More databases exist than you ever imagined! Page created and maintained by Amos Bairoch.

Computational services at EMBL
http://WWW.EMBL-Heidelberg.de/Services/
A wide range of sequence analysis tools and links from EMBL scientists.

More links at the Bio-Web
http://cellbiol.com/Tools.html
The Bio-Web sequence analysis tools page

7

Theoretical Aspects of Sequence Alignments

Barbara Brannetti and Allegra Via

Contents

From: *The Internet for Cell and Molecular Biologists: Current Applications and Future Potential*
ISBN 1-898486-32-8 © 2002 Horizon Scientific Press, Wymondham, UK

Abstract

This chapter is dedicated to the theoretical aspects of the analysis of nucleic and amino acid sequences. It consists of two main sections: a 'pair-wise alignments' part (section 7.1) and a 'multiple alignments' part (section 7.2) where the reader can find an outline of the concepts underlying pair-wise and multiple (DNA and protein) sequence alignments together with a theoretical discussion of the principles regulating the most important algorithms for sequence analysis. This is not essential for the comprehension and full usage of chapter 8 and chapter 9, but may help the reader who wishes to get a deeper view of the subject.

Therefore those who are interested in the practical use of sequence databases and programs for sequence analysis can skip this chapter and go directly to chapter 8 or chapter 9.

7.1 Pairwise alignments

7.1.1 Alignments

Sequence alignment provides a powerful tool for the inference of homology from two or more sequences that display sufficient similarity. These two terms, homology and similarity, are often used as synonyms, but they are not. Homology has the meaning of 'sharing a common evolutionary history'. In other words two sequences are homologous if they share a common ancestor sequence. Thus, sequences are either homologous or not. Similarity is an observed quantity, that may be expressed in percent of identity, for example sequences may have 30 %, 40 % identity. The changes occurring can be categorized as substitutions, insertions and deletions. Aligned residues that are not identical can represent substitutions with respect to a supposed ancestor sequence, while regions in which residues of one sequence corresponding to nothing in the other sequence can be an insertion into the first sequence or a deletion in the second one. The best alignment cannot be unambiguously established; this is due to the computational complexity of the problem and to the still limited knowledge of the mutation frequency during evolution.

7.1.2 Global and local alignment

Sequences can be homologous over their entire length, in which case it is possible to align the sequences with computational procedures based on the Needleman-Wunsch (Needleman and Wunsch, 1970) global alignment algorithm. In the formulation of this algorithm the best alignment between two sequences has to be the one covering both sequences from start to end. In other situations the sequence similarity is restricted to limited regions of the sequences. Many proteins consist of a combination of 'modules', such as the adapter proteins, that have been shuffled during evolution. A given module may be repeated within a sequence and even found in several proteins with different function. In that case, methods based on the Smith-Waterman (Smith and Waterman, 1981) algorithm perform better. These methods are designed to find the regions of highest similarity within two sequences.

The BLAST (Basic Local Alignment Search Tool) and FASTA algorithms are two of these methods. It is important to remember that these algorithms, the global and the local, report the best alignment that they can evaluate even if it has no biological meaning. These algorithms work on models that try to explain the molecular mechanisms by which sequences evolve. Thus, it is better not to look only at the best alignment found.

7.1.3 Substitutions

The probability of substitution depends on factors such as the genetic code and the phenotypic effect of the mutation for amino acid residues. At the DNA level, the probability of mutation depends on the nature of the base. Transitions (purines to purines, pyrimidines to pyrimidines) are more frequent than transversions (purines to pyrimidines and *vice versa*), and depend also on the neighbouring bases (the C in the CG pairs of nucleotides mutates with high frequency in vertebrates) (Bains, 1992; Hess *et al*, 1994). When an alignment is performed with global or local algorithms, residue pairs are evaluated for their similarity and identical residues should have higher scores than different ones, and conservative substitutions a higher score than non-conservative changes. Thus the scoring system has to access a range of values, i.e. different substitution matrices, collecting sets of values for comparison of very similar sequences, such as those from rat and mouse organisms, and more distantly related ones, such as those from human and yeast organisms.

A substitution matrix is a table where a value is calculated for each possible residue-residue pair (Figure 7.1 and Figure 7.2). Generally, pairs of identical residues are associated with the highest values in the substitution matrix. Residue pairs sharing similar chemico-physical features (e.g. arginine and lysine) are generally associated with lower values. The lowest values are found associated with residue pairs that are very different in volume and side-chain polarity (e.g. tryptophan and glycine).

Different substitution matrices can be used in different similarity searches and the choice can strongly influence the outcome of an analysis. The most widely used are the PAM and BLOSUM substitution matrices.

The PAM matrices are based on the point-accepted mutation (PAM) model of evolution described by Dayhoff and co-workers (Figure 7.1) (Dayhoff et al., 1978). The evolutionary divergence between two sequences is measured in PAM units, 1 PAM means that 1% of amino acid positions have been changed. This does not mean that after 100 PAMs all the residues have changed; some will have mutated several times, maybe returning to their original state, others will not. Closely related sequences were chosen to evaluate the mutation frequencies corresponding to 1 PAM distance (for details see Dayhoff et al., 1978). The collected mutation data were extrapolated to other PAM distances in order to produce a set of matrices to compare similar as well as highly divergent sequences. Thus, several PAM matrices were calculated that are identified by increasing numbers, such as PAM80, PAM120, PAM200 and PAM250, indicating increasing evolutionary distances.

The BLOSUM matrices (Henikoff and Henikoff, 1992) have been constructed in a similar way (Figure 7.2). The basic idea was to get a better measure of the differences between distantly related proteins. To collect the mutation frequencies of the residues, the data set was derived from the BLOCKS database, (BLOSUM stands for **BLOcks SUbstitution Matrices**) which contains local multiple

```
#
# This matrix was produced by "pam" Version 1.0.6 [ 28-Jul-
93]
#
# PAM 120 substitution matrix,  scale = ln(2)/2 = 0.346574
#
# Expected score = -1.64, Entropy = 0.979 bits
#
# Lowest score = -8, Highest score = 12
#
    A  R  N  D  C  Q  E  G  H  I  L  K  M  F  P  S  T  W  Y  V  B  Z  X  *
A   3 -3 -1  0 -3 -1  0  1 -3 -1 -3 -2 -2 -4  1  1  1 -7 -4  0  0 -1 -1 -8
R  -3  6 -1 -3 -4  1 -3 -4  1 -2 -4  2 -1 -5 -1 -1 -2  1 -5 -3 -2 -1 -2 -8
N  -1 -1  4  2 -5  0  1  0  2 -2 -4  1 -3 -4 -2  1  0 -4 -2 -3  3  0 -1 -8
D   0 -3  2  5 -7  1  3  0  0 -3 -5 -1 -4 -7 -3  0 -1 -8 -5 -3  4  3 -2 -8
C  -3 -4 -5 -7  9 -7 -7 -4 -4 -3 -7 -7 -6 -6 -4  0 -3 -8 -1 -3 -6 -7 -4 -8
Q  -1  1  0  1 -7  6  2 -3  3 -3 -2  0 -1 -6  0 -2 -2 -6 -5 -3  0  4 -1 -8
E   0 -3  1  3 -7  2  5 -1 -1 -3 -4 -1 -3 -7 -2 -1 -2 -8 -5 -3  3  4 -1 -8
G   1 -4  0  0 -4 -3 -1  5 -4 -4 -5 -3 -4 -5 -2  1 -1 -8 -6 -2  0 -2 -2 -8
H  -3  1  2  0 -4  3 -1 -4  7 -4 -3 -2 -4 -3 -1 -2 -3 -3 -1 -3  1  1 -2 -8
I  -1 -2 -2 -3 -3 -3 -3 -4 -4  6  1 -3  1  0 -3 -2  0 -6 -2  3 -3 -3 -1 -8
L  -3 -4 -4 -5 -7 -2 -4 -5 -3  1  5 -4  3  0 -3 -4 -3 -3 -2  1 -4 -3 -2 -8
K  -2  2  1 -1 -7  0 -1 -3 -2 -3 -4  5  0 -7 -2 -1 -1 -5 -5 -4  0 -1 -2 -8
M  -2 -1 -3 -4 -6 -1 -3 -4 -4  1  3  0  8 -1 -3 -2 -1 -6 -4  1 -4 -2 -2 -8
F  -4 -5 -4 -7 -6 -6 -7 -5 -3  0  0 -7 -1  8 -5 -3 -4 -1  4 -3 -5 -6 -3 -8
P   1 -1 -2 -3 -4  0 -2 -2 -1 -3 -3 -2 -3 -5  6  1 -1 -7 -6 -2 -2 -1 -2 -8
S   1 -1  1  0  0 -2 -1  1 -2 -2 -4 -1 -2 -3  1  3  2 -2 -3 -2  0 -1 -1 -8
T   1 -2  0 -1 -3 -2 -2 -1 -3  0 -3 -1 -1 -4 -1  2  4 -6 -3  0  0 -2 -1 -8
W  -7  1 -4 -8 -8 -6 -8 -8 -3 -6 -3 -5 -6 -1 -7 -2 -6 12 -2 -8 -6 -7 -5 -8
Y  -4 -5 -2 -5 -1 -5 -5 -6 -1 -2 -2 -5 -4  4 -6 -3 -3 -2  8 -3 -3 -5 -3 -8
V   0 -3 -3 -3 -3 -3 -3 -2 -3  3  1 -4  1 -3 -2 -2  0 -8 -3  5 -3 -3 -1 -8
B   0 -2  3  4 -6  0  3  0  1 -3 -4  0 -4 -5 -2  0  0 -6 -3 -3  4  2 -1 -8
Z  -1 -1  0  3 -7  4  4 -2  1 -3 -3 -1 -2 -6 -1 -1 -2 -7 -5 -3  2  4 -1 -8
X  -1 -2 -1 -2 -4 -1 -1 -2 -2 -1 -2 -2 -2 -3 -2 -1 -1 -5 -3 -1 -1 -1 -2 -8
*  -8 -8 -8 -8 -8 -8 -8 -8 -8 -8 -8 -8 -8 -8 -8 -8 -8 -8 -8 -8 -8 -8 -8  1
```

Figure 7.1. Substitution matrix table: PAM matrix.

alignments, named 'blocks', of distantly related sequences. As for the PAM matrices, a series of matrices was calculated. Here the numbers (BLOSUM80, BLOSUM62, BLOSUM45..) refer to the level of identity of aligned sequences of the 'block' used to built the relative matrix. For example, with BLOSUM62 sequences sharing at least 62% identity are merged together. Thus BLOSUM substitution matrices with higher numbers indicate smaller evolutionary distances, and *vice versa*.

Substitution matrices are also available for DNA sequences. These matrices contain information about the occurrence of transitions and transversions (Li and Graur 1991), and may be used for local or global alignment and for database searching. The programs based on the Needleman-Wunsch global alignment algorithm use a positive score for matches and zero for mismatches. The ones based on the Smith-Watermann local alignment algorithm define the end of a region of similarity using a negative score for mismatching. A series of PAM nucleic acid substitution matrices were developed by States and co-workers (States *et al.*, 1991) to improve the sensitivity of similarity searches in nucleic acid databases. They are also used to score the nucleic acid alignments. These matrices take into account different

```
#   Matrix made by matblas from blosum62.iij
#   * column uses minimum score
#   BLOSUM Clustered Scoring Matrix in 1/2 Bit Units
#   Blocks Database = /data/blocks_5.0/blocks.dat
#   Cluster Percentage: >= 62
#   Entropy =   0.6979, Expected =  -0.5209
    A  R  N  D  C  Q  E  G  H  I  L  K  M  F  P  S  T  W  Y  V  B  Z  X  *
A   4 -1 -2 -2  0 -1 -1  0 -2 -1 -1 -1 -1 -2 -1  1  0 -3 -2  0 -2 -1  0 -4
R  -1  5  0 -2 -3  1  0 -2  0 -3 -2  2 -1 -3 -2 -1 -1 -3 -2 -3 -1  0 -1 -4
N  -2  0  6  1 -3  0  0  0  1 -3 -3  0 -2 -3 -2  1  0 -4 -2 -3  3  0 -1 -4
D  -2 -2  1  6 -3  0  2 -1 -1 -3 -4 -1 -3 -3 -1  0 -1 -4 -3 -3  4  1 -1 -4
C   0 -3 -3 -3  9 -3 -4 -3 -3 -1 -1 -3 -1 -2 -3 -1 -1 -2 -2 -1 -3 -3 -2 -4
Q  -1  1  0  0 -3  5  2 -2  0 -3 -2  1  0 -3 -1  0 -1 -2 -1 -2  0  3 -1 -4
E  -1  0  0  2 -4  2  5 -2  0 -3 -3  1 -2 -3 -1  0 -1 -3 -2 -2  1  4 -1 -4
G   0 -2  0 -1 -3 -2 -2  6 -2 -4 -4 -2 -3 -3 -2  0 -2 -2 -3 -3 -1 -2 -1 -4
H  -2  0  1 -1 -3  0  0 -2  8 -3 -3 -1 -2 -1 -2 -1 -2 -2  2 -3  0  0 -1 -4
I  -1 -3 -3 -3 -1 -3 -3 -4 -3  4  2 -3  1  0 -3 -2 -1 -3 -1  3 -3 -3 -1 -4
L  -1 -2 -3 -4 -1 -2 -3 -4 -3  2  4 -2  2  0 -3 -2 -1 -2 -1  1 -4 -3 -1 -4
K  -1  2  0 -1 -3  1  1 -2 -1 -3 -2  5 -1 -3 -1  0 -1 -3 -2 -2  0  1 -1 -4
M  -1 -1 -2 -3 -1  0 -2 -3 -2  1  2 -1  5  0 -2 -1 -1 -1 -1  1 -3 -1 -1 -4
F  -2 -3 -3 -3 -2 -3 -3 -3 -1  0  0 -3  0  6 -4 -2 -2  1  3 -1 -3 -3 -1 -4
P  -1 -2 -2 -1 -3 -1 -1 -2 -2 -3 -3 -1 -2 -4  7 -1 -1 -4 -3 -2 -2 -1 -2 -4
S   1 -1  1  0 -1  0  0  0 -1 -2 -2  0 -1 -2 -1  4  1 -3 -2 -2  0  0  0 -4
T   0 -1  0 -1 -1 -1 -1 -2 -2 -1 -1 -1 -1 -2 -1  1  5 -2 -2  0 -1 -1  0 -4
W  -3 -3 -4 -4 -2 -2 -3 -2 -2 -3 -2 -3 -1  1 -4 -3 -2 11  2 -3 -4 -3 -2 -4
Y  -2 -2 -2 -3 -2 -1 -2 -3  2 -1 -1 -2 -1  3 -3 -2 -2  2  7 -1 -3 -2 -1 -4
V   0 -3 -3 -3 -1 -2  2 -3 -3  3  1 -2  1 -1 -2 -2  0 -3 -1  4 -3 -2 -1 -4
B  -2 -1  3  4 -3  0  1 -1  0 -3 -4  0 -3 -3 -2  0 -1 -4 -3 -3  4  1 -1 -4
Z  -1  0  0  1 -3  3  4 -2  0 -3 -3  1 -1 -3 -1  0 -1 -3 -2 -2  1  4 -1 -4
X   0 -1 -1 -1 -2 -1 -1 -1 -1 -1 -1 -1 -1 -2  0  0 -2 -1 -1 -1 -1 -1 -1 -4
*  -4 -4 -4 -4 -4 -4 -4 -4 -4 -4 -4 -4 -4 -4 -4 -4 -4 -4 -4 -4 -4 -4 -4  1
```

Figure 7.2. Substitution matrix table: BLOSUM matrix.

levels of sequence divergence and different rates of transversions and transitions.

7.1.4 Insertions and deletions

Performing an alignment means aligning the residues of the sequences and allowing gaps to introduce insertions and/or deletions. The scoring system of the alignment programs considers two terms for the gap penalty, the cost to open the gap, namely the gap open penalty, and the cost to extend the gap, namely the gap extension penalty. The choice of the appropriate values for these parameters is empirical; most programs use as default values those derived from the analysis of globular proteins, but this is not optimal for other types of sequences. The selection of suitable values is very important particularly during a database search for homologous proteins of the query sequence, to distinguish them from the non-homologous ones. Choosing gap penalties that are too small can increase the numbers of non-homologous sequences with high similarity scores to the query sequence, thus the matches to the homologous sequences can be shuffled and obscured by matches to the non-homologous one (Vingron *et al.*, 1994).

7.1.5 Statistical significance of alignments

In order to provide evidence of homology in a given alignment it is very important to know how significant is its score, calculated as the sum of substitution and gap scores. This is achieved by evaluating the probability of obtaining the same score by chance. The distribution of similarity scores observed for the global alignments cannot be described by a mathematical theory. Thus, one of the common methods used is to compare observed alignment scores with the ones coming from alignments obtained from random sequences with the same length and composition of the sequences of interest (Fitch, 1983). For local alignments, Karlin and Altschul (Karlin and Altschul, 1990; Karlin and Altschul, 1993) have developed a statistical model to evaluate the significance of local un-gapped alignments. These types of local alignments are called high-scoring segment pairs (HSPs), and can be found by imposing high values for the gap penalties. With this method it is possible to evaluate the expected distribution of random HSPs scores and calculate for a given alignment score, S, a p value which gives the probability that an alignment with that score or better could be due to chance. The most highly significant p values will be those close to 0. A related quantity is the E value, that is the expected number of alignments due only by chance with a score at least equal to S; the lower the E value, the more significant the score. The p and E values embody different ways of representing the significance of an alignment. The same statistical model applied to the HSPs can be extended to the un-gapped alignments. This is strongly supported by computational experiments and analytical results (Arratia *et al.*, 1994).

7.2 Multiple alignments

7.2.1 Introduction

In order to build a multiple alignment of protein sequences (Figure 7.3), residues deriving from a common ancestor have to be placed in the same columns. To this aim, if needed, insertions, deletions and substitutions can be used (as discussed for pair-wise alignments).

As for a pair-wise alignment (see Section 7.1), the best multiple alignment should reflect the most likely evolutionary history of the aligned sequences. On this basis, regions conserved in different protein sequences are generally structurally or functionally important. In this section we report the information contained in a multiple sequence alignment (MSA) and the most widely used tools to build the alignment.

7.2.2 Multiple alignments: why do we need them?

Since residues conserved during evolution usually correspond to structurally or functionally important protein regions, multiple alignments can supply important information for function and/or structure (see Chapter 10). Moreover, patterns of conserved residues

```
90
                1               15 16            30 31            45 46             60 61             75 76
HBA_HUMAN   -VLSPADKTNVKAAW GKVGAHAGEYGAEAL ERMFLSFPTTKTYFP HF------DLSHGSA QVKGHGKKVADALTN AVAHVDDMPNALSAL  83
HBA_PHYCA   -VLSPADKTNVKAAW AKVGNHAADFGAEAL ERMFMSFPSTKTYFS HF------DLGHNST QVKGHGKKVADALTK AVGHLDTLPDALSDL  83
HBB_HUMAN   VHLTPEEKSAVTALW GKVN--VDEVGGEAL GRLLVVYPWTQRFFE SFGDLSTPDAVMGNP KVKAHGKKVLGAFSD GLAHLDNLKGTFATL  88
HBB_PHYCA   VHLTGEEKSGLTALW AKVN--VEEIGGEAL GRLLVVYPWTQRFFE HFGDLSTADAVMKN? KVKKHGQKVLASFGE GLKHLDNLKGTFATL  88
HBBF_BOVIN  -MLSAEEKAAVTSLF AKVK--VDEVGGEAL GRLLVVYPWTQRFFE SFGDLSSADAILGNP KVKAHGKKVLDSFCE GLKQLDDLKGAFASL  87
BAHG_VITST  -MLDQQTINIIKATV PVLKEHGVTITTTFY KNLFAKHPEVRPLFD MG------RQESL  EQPKALAMTVLAAAQ NIENLPAILPAVKKI  81

                91              105 106           120 121          135 136           150 151           165 166
HBA_HUMAN   SDLHAHKLRVDPVNF KLLSHCLLVTLAAHL PAEFTPAVHASLDKF LASVSTVLTSKYR-- -------  141
HBA_PHYCA   SDLHAHKLRVDPVNF KLLSHCLLVTLAAHL PGDFTPSVHASLDKF LASVSTVLTSKYR-- -------  141
HBB_HUMAN   SELHCDKLHVDPENF RLLGNVLVCVLAHHF GKEFTPVQAA?QKV  VAGVANALAHKYH-- -------  146
HBB_PHYCA   SELHCDKLHVDPENF RLLGNVLVVVLARHF GKEFTPELQTAYQKV VAGVANALAHKYH-- -------  146
HBBF_BOVIN  SELHCDKLHVDPENF RLLGNVLVVVLARRF GSEFSPELQASFQKV VTGVANALAHRYH-- -------  145
BAHG_VITST  AVKHCQAG-VAAAHY PIVGQELLGAIKEVL GDAATDILDAWGKA  YGVIADVFIQVEADL YAQAVE   146
180
```

Figure 7.3. The multiple alignment of six sequences of hemoglobin: the hemoglobin alpha and beta chain of human (HBA_HUMAN, HBB_HUMAN), the hemoglobin alpha and beta chain of sperm whale (HBA_PHYCA, HBB_PHYCA) the fetal hemoglobin beta chain of bovin (HBBF_BOVIN) and the bacterial hemoglobin of Vitreoscilla stercoraria. The multiple alignment has been obtained with the program ClustalW (see chapter 9). It is interesting to notice that the hemoglobin alpha chains of human and sperm whale are more similar than the hemoglobin alpha and beta chain of the same organism. Moreover, the growth of the evolutionary distance between sequences corresponds to a decreasing number of conserved residues in the columns of the multiple alignment.

in individual columns of a multiple alignment may give important information on the natural constraints acting on a molecule.

Therefore the alignment of a new sequence to a family of already characterized proteins may allow the identification of functionally or structurally important features with good reliability.

For example, well conserved regions in a multiple alignment usually represent secondary structure elements, whereas regions characterized by indels (i.e. insertions and deletions) identify loops between these elements.

Moreover, multiple alignments can be useful for inferring homology among sequences (see Section 7.1.1): in many cases, similarities that do not seem significant (in a pair-wise alignment) may be shown to be significant in a multiple alignment because of conservation across distantly related sequences.

The probability of detecting weak similarities in a sequences database can be strongly increased using *position–specific scoring matrices* (PSSMs). PSSMs are obtained from multiple alignments and consist of columns of weights for each residue according to its degree of conservation. A simple PSSM has 20 columns and as many rows as there are positions in the alignment. Methods based on PSSMs are described in chapter 9.

Another important application of multiple alignments consists of retracing evolutionary relationships between sequences: phylogenetic trees rely on multiple alignments for inferring mutation events that occurred during evolution and therefore reconstruction of the history of species.

7.2.3 Global and local alignments

As for pair-wise alignments (see Section 7.1), in many cases similarity is limited to a region (or a few regions) of the analysed sequences. In such cases it is not possible to construct a multiple alignment over the entire length of the sequences and only local alignments across homologous modules can be built. These local alignments can be very interesting for the study of the evolution of protein modules.

The general theory of global and local alignments is the same as for pair-wise sequence alignments. See Section 7.1.2.

7.2.4 Substitutions, deletions and insertion

As stated above, the best multiple alignment should reflect the most likely evolutionary scenario. To this aim, both computational and non-computational methods that are used to perform alignments have to deal with substitutions and indels by the use of a suitable scoring procedure. The treatment of this subject is the same as for pair-wise sequence alignments. See Section 7.1.4.

7.2.5 How do we obtain a multiple alignment?

For the computation of a multiple alignment, first of all homologous sequences have to be identified. This can be done by means of pair-wise alignments and database searches as described above (see Section 7.1). Once a set of homologous sequences is obtained, the user should choose the most appropriate multiple alignment method.

Several computational tools have been developed to solve the problem of aligning a set of homologous sequences. Among them it is possible to distinguish methods that – for a given set of substitution matrices and gap penalty parameters – find the optimal alignment and methods based on heuristic algorithms. None are ideal but some, especially those that use heuristic algorithms, are very useful in practice.

Methods for 'optimal' alignments have the strong constraint that they require a lot of computer time and memory. As a consequence they can be used only to align a small number (10 - 15) of short sequences (length ~100 residues). They are based on the concept of **alignment score** (or **cost**) **scheme**. The measure of the cost of a column of an alignment takes into account the weight of substitutions (see section 7.1.3) and the gap penalty parameters. The cost of a multiple alignment can be calculated in different ways. The simplest model considers each column independently of its context and calculates the cost of a multiple alignment as the sum of the costs of its columns. In a more realistic model, it is reasonable to consider a cost for the gap opening plus a smaller cost for the gap length. While this calculation is quite linear for a pair-wise alignment, it is more complex for a multiple alignment of n $(n>2)$ sequences and consists essentially of calculating the multiple alignment cost from all the $n(n-1)/2$ pair-wise alignment costs. In this approach it is also possible to consider the bias represented by groups of sequences that are over- or under-represented in a family, by conveniently weighting each sequence (less weight to sequences that are over-represented and *vice versa*).

The alignment with the minimal cost (or maximal score) represents the optimal one. However, only the proper choice of parameters (substitution matrix and indels penalties) guarantees that the optimal alignment corresponds to the biologically correct alignment.

Heuristic algorithms for the construction of multiple alignments can be divided in three main subgroups:

1.   Algorithms based on progressive pair-wise alignment. These are the most commonly used to align sequences (see Chapter 9).

2.   Algorithms that produce global alignments based on local alignments.

3.   Algorithms that produce local alignments.

1. Algorithms based on progressive pair-wise alignment

This approach represents a good trade-off between the speed and the possibility of obtaining an alignment close enough to the optimal one. In any case, as for every multiple alignment obtained by means of computational methods, the resulting alignment *always* has to be examined to verify whether changes that account for biological factors are needed.

The progressive alignment procedure consists essentially of three steps. Different methods relying on this procedure differ in at least one of these steps in a way that is too detailed/specific to describe in this context.

First, sequence pairs are aligned and the cost (or *distance*) of the alignments is calculated (using dynamic programming or heuristic algorithms). Second, they are listed in a table (*distance matrix*) and this table is used to obtain a tree that reflects distances/similarities between sequences. Finally, following the tree (see Figure 7.4) from the most distant leaves towards the root, pairs of sequences are aligned, then pair-wise alignments are progressively aligned between them and to other sequences at the same branching level of the tree.

The general procedure of the progressive pair-wise alignment of five sequences s1,s2,s3,s4,s5 is described schematically in Figure 7.4.

2. Algorithms that produce global alignments based on local alignments

Another group of methods uses heuristic algorithms to obtain global alignments based on local alignments (block-based global alignment). In some cases the sequences to be aligned show local alignments (blocks) interspersed in large non-conserved regions. In such cases, the blocks can be used as *anchors* in order to compute a multiple global alignment, using classical procedures to align regions between blocks.

distance matrix

	S1	S2	S3	S4	S5
S1	0.00	0.24	0.11	0.31	0.35
S2		0.00	0.29	0.08	0.33
S3			0.00	0.28	0.30
S4				0.00	0.34
S5					0.00

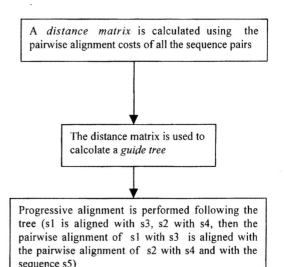

A *distance matrix* is calculated using the pairwise alignment costs of all the sequence pairs

The distance matrix is used to calcolate a *guide tree*

Progressive alignment is performed following the tree (s1 is aligned with s3, s2 with s4, then the pairwise alignment of s1 with s3 is aligned with the pairwise alignment of s2 with s4 and with the sequence s5)

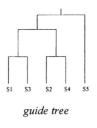

S1 S3   S2 S4   S5

guide tree

Figure 7.4. The general procedure of the progressive pair-wise alignment of the five sequences s1,s2,s3,s4,s5.

3. Algorithms that produce local alignments

Sometimes it is not possible to find a global alignment starting from a set of sequences. However, these sequences may share similar modules (sometimes duplicated and/or occurring in different relative positions) separated by non-conserved regions of variable length that cannot be aligned. In other words, the sequences are not related over their entire length, but display local regions of homology. In such cases it may be possible to use heuristic algorithms to calculate local alignments.

In the list of methods for the construction of multiple alignments, those based on HMMs (Hidden Markov Models) and on genetic algorithms should also be considered. The general theory of HMM and genetic algorithms will not be discussed in this introduction, since their treatment requires a firm mathematical background, however in section 9.4 (Hidden Markov Models) the practical computing of sequence alignments by means of HMM will be described briefly, and some literature references and WWW sites on the subject are provided.

7.2.6 Gene prediction and pattern matching

Searching for common motifs in a set of unrelated sequences is a subject somewhat related to that of multiple alignments. A motif may be considered as a short sequence fragment conserved in originally homologous sequences that have then diverged.

The identification of special regions (such as promoters or splicing sites for DNA sequences and functional sites for protein sequences) of a newly-determined sequence may help the biologist in its characterization. Both for DNA and protein sequences the 'protocol' consists of scanning databases of 'motifs' with a query sequence. However, the type of database and programs used in the case of nucleic acid sequences is rather different from those used for amino acid sequences. Therefore they will be discussed separately in chapter 8 (gene prediction) and 9 (pattern matching), respectively.

7.3 References

Arratia, R. and Waterman, M.S. 1994. A phase transition for the score in matching random sequences allowing deletions. Ann. Appl. Prob. 4: 200-225.

Bains W. 1992. Local sequence dependence of rate of base replacement in mammals. Mutat. Res. 267: 43-54.

Dayhoff, M. O., Schwartz, R. M. and Orcutt, B. C. 1978. A model of evolutionary change in proteins. matrices for detecting distant relationships. In Atlas of protein sequence and structure, (Dayhoff, M. O., ed.), vol. 5, pp. 345-358. National biomedical research foundation Washington DC.

Fitch, WM. 1983. Random sequences. J. Mol. Biol. 163: 171-176.

Henikoff, S. and Henikoff, J. G. 1992. Amino acid substitution matrices from protein blocks. Proc. Natl. Acad. Sci. USA, 89, 10915-10919.

Hess S.T., Blake J.D., Blake R.D. 1994. Wide variations in neighbor-dependent substitution rates. J.Mol.Biol. 236: 1022-33.

Karlin, S. and Altschul, S.F. 1990. Methods for assessing the statistical significance of molecular sequence features by using general scoring schemes. Proc. Natl. Acad. Sci. USA. 87: 2264 - 2268.

Karlin, S. and Altschul, S.F. 1993. Applications and statistics for multiple high-scoring segments in molecular sequences. Proc. Natl. Acad. Sci. USA. 90: 5873 - 5877.

Li, W., and Graur, D. 1991. Fundamentals of molecular evolution. Sinauer Associates, Inc. Sunderland Mass.

Needleman, S. B. and Wunsch, C. D. 1970. A general method applicable to the search for similarities in the amino acid sequence of two proteins. J. Mol. Biol. 48: 443-453.

Smith, T. F. and Waterman, M. S. 1981. Identification of common molecular subsequences. J. Mol. Biol. 147: 195-197.

States, D.J., Gish, W., Altschul, S.F. 1991. Improved sensitivity of nucleic acid database search using application-specific scoring matrices. Methods: A companion to Meth. Enzymol. 3: 66 - 77.

Vingron, M. and Waterman, M.S. 1994. Sequence alignment and penalty choice. J. Mol. Biol. 235: 1 - 12.

8

Analyse DNA Sequences With Your Browser

Barbara Brannetti

Contents

From: *The Internet for Cell and Molecular Biologists: Current Applications and Future Potential*
ISBN 1-898486-32-8 © 2002 Horizon Scientific Press, Wymondham, UK

Abstract

The enormous amount of data coming from the various genome projects is stored within biological databases. Different tools have been developed both to search within the databases and to analyse and annotate the contained data. The aim of this chapter is to describe the more useful and used nucleic acid databases and to introduce the tools developed to analyse nucleic acid sequences. It is organized into three main sections. The first (8.1) deals with a description of the Genbank database, with details of the structure of the files containing sequence data together with some annotation. The second section (8.2) provides a user-friendly description of tools (FASTA and BLAST) for the comparison of a query sequence with a nucleic acid database. A detailed description of the more useful tools available for gene structure prediction is reported in section 8.3. The prediction of functional sites in a raw genomic sequence is still a hot research topic (cf. Fortna and Gardiner, 2001) and no easy solution and completely reliable tool can be presented so far. We suggest therefore trying different tools in order to compare the different predictions and identify the method that seems to be more reliable for the reader's specific problem.

8.1 GenBank Database

In this section the most complete collection of DNA sequence data available, the GenBank database, is described. But it is important to bear in mind that GenBank is part of a community of databases that also includes two protein sequence databases, the Swiss-Prot (http://www.expasy.ch/sprot), the Protein Information Resource (PIR, http://pir.georgetown.edu/) database, and the Protein Data Bank (PDB, http://www.rcsb.org/pdb/), that is the repository of the protein three-dimensional structures.

GenBank was built by the National Center for Biotechnology Information (NCBI) at NIH in Bethesda, and maintained in cooperation with two partners, the DNA Database of Japan (DDBJ) at Mishima, and the European Molecular Biology Laboratory (EMBL) nucleotide database maintained by the European Bioinformatics Institute (EBI) at Hinxton.

The three centres work in close collaboration; they are distinct points for submitting data but data exchange between them occurs daily. The database may be accessed at NCBI at http://www.ncbi. nlm.nih.gov/Genbank/, at EBI at http://www.ebi.ac.uk/embl/, and at DDBJ at http://www.ddbj.nig.ac.jp/.

8.1.1 Description of GenBank database records

The representation of each record has a specific format, called flatfile. The GenBank flatfile (GBFF) and the DDBJ flatfile are identical; the EMBL flatfile uses line type prefixes, i.e the line starting with 'FT' is equivalent to the line beginning 'FEATURES' in the GBFF and DDBJ flatfile.

The flatfile can be divided into three parts: the header, the features and the sequence.

The header (Figure 8.1) is the only section that can vary among the database flatfiles. Here follows a description of its structure.

LOCUS:
Is the first line of the file. It consists of a *locus name*, a term originally designed to help in grouping entries with similar sequences. The rule now applied to assign this term is that it must be unique, and it is built with the first letter of the genus and species name followed by the accession number or just the accession number.

The number of nucleotide base pairs follows the locus name, then the type of sequence is reported. The following term refers to the GenBank division (there are actually 16 divisions) see Table 8.1.

- The date in LOCUS line refers to the last modification. Sometimes the release date is reported.
- The next line is the DEFINITION, a brief description of the sequence containing the gene or protein name, the source organism and the function of the sequence if known.
- The ACCESSION number follows. It is a unique identifier, a combination of one or two letters with numbers. It does not change even if the record is modified.

```
LOCUS       HSCSKPTK 5653 bp DNA PRI 18-OCT-1999
DEFINITION  Homo sapiens CSK gene for protein tyrosine kinase.
ACCESSION   X74765
VERSION     X74765.1   GI:402582
KEYWORDS    c-src tyrosine kinase; CSK gene; protein tyrosine kinase.
SOURCE      human.
  ORGANISM  Homo sapiens
            Eukaryota; Metazoa; Chordata; Craniata; Vertebrata;
Euteleostomi; Mammalia; Eutheria; Primates; Catarrhini; Hominidae;
Homo.
REFERENCE   1  (bases 1 to 5653)
  AUTHORS   Brauninger,A., Karn,T., Strebhardt,K. and Rubsamen-
Waigmann,H.
  TITLE     Characterization of the human CSK locus
  JOURNAL   Oncogene 8 (5), 1365-1369 (1993)
  MEDLINE   93241739
REFERENCE   2  (bases 1 to 5653)
  AUTHORS   Karn,T.
  TITLE     Direct Submission
  JOURNAL   Submitted (19-AUG-1993) T. Karn, Chemotherapeutisches
Forschungsinstitut, Georg-Speyer-Haus, Paul Ehrlich Strasse 42-44,
D60596 Frankfurt, FRG
```

Figure 8.1. The header of a GenBank database file.

Table 8.1. GenBank divisions.

PRI	Primate sequences
ROD	Rodent sequences
MAM	Other mammalian sequences
VRT	Other vertebrate seqeunces
INV	Invertebrate sequences
PLN	Plant,fungal and algal sequences
BCT	Bacterial sequences
VRL	Viral sequences
PHG	Bacteriophage sequences
SYN	Synthetic sequences
UNA	Unannotated sequences
EST	EST sequences
PAT	Patent sequences
STS	Sequence Tagged Site sequences
GSS	Genome Survey Sequences
HTG	High Throughput Genomic sequences

- VERSION is the line reporting the record version, if the record changes the version is increased, i.e AF001863.1->AF001863.2
- The GI number is the GenInfo identifier, a sequence identification number, assigned to each record. If there is a change in a record, a new GI is assigned and the version number increases. Version number and GI system run in parallel.
- The KEYWORDS is a line with words describing the sequence. This field is not under the control of a vocabulary, and the keywords are no longer included in new records.
- The SOURCE line contains information about the source organism.
- The ORGANISM term reports the formal scientific name for the source, genus and species, and it is based on the phylogenetic classification scheme of the NCBI Taxonomy Database.
- The last part of the HEADER reports the published articles dealing with the data entry. The references are automatically sorted, and the oldest one is the first listed. In the reference list, the following is reported: the authors list, the title of the published work or the tentative title for the unpublished one, the MEDLINE and PUBMED identifiers, that link the sequence to the corresponding MEDLINE record and to the literature database of PUBMED, and the contact information of the submitter/s.

The second section of the GenBank flatfile (Figure 8.2) is dedicated to the information about the gene and the gene product and it starts with the FEATURES term. *Source* reports data about the length of the sequence, the scientific name and the taxonomy identification number of the source organism. It can also include information about the clone, the tissue type, the cell line and the strain. The CDS (Coding Sequence) feature follows. It reports, if known, the region of the sequence that codes for the protein, with the translation. A protein sequence identification number is associated with this coding sequence in the accession version format. If there is any change in the sequence, the

```
FEATURES                Location/Qualifiers
    source              1..5653
                        /organism="Homo sapiens"
                        /db_xref="taxon:9606"
                        /clone="lambda EMBL4-CSK1"
                        /clone_lib="genomic lambda EMBL4"
    mRNA
    join(81..160,463..576,674..786,1120..1339,2260..2353,
    2529..2594,2671..2770,2860..2950,3371..3444,3544..3739,
    3846..3932,4180..5045)
                        /gene="CSK"
                        /product="protein tyrosine kinase"
    gene                81..5045
                        /gene="CSK"
    prim_transcript 81..5045
                        /gene="CSK"
    exon                81..160
                        /gene="CSK"
                        /number=1
    CDS
    join(146..160,463..576,674..786,1120..1339,2260..2353,
    2529..2594,2671..2770,2860..2950,3371..3444,3544..3739,
    3846..3932,4180..4362)
                        /gene="CSK"
                        /codon_start=1
                        /product="protein tyrosine kinase"
                        /protein_id="CAB58562.1"
                        /db_xref="GI:6077093"
/translation="MSAIQAAWPSGTECIAKYNFHGTAEQDLPFCKGDVLTIVAVTKD
PNWYKAKNKVGREGIIPANYVQKREGVKAGTKLSLMPWFHGKITREQAERLLYPPETG
LFLVRESTNYPGDYTLCVSCDGKVEHYRIMYHASKLSIDEEVYFENLMQLVEHYTSDA
DGLCTRLIKPKVMEGTVAAQDEFYRSGWALNMKELKLLQTIGKGEFGDVMLGDYRGNK
VAVKCIKNDATAQAFLAEASVMTQLRHSNLVQLLGVIVEEKGGLYIVTEYMAKGSLVD
YLRSRGRSVLGGDCLLKFSLDVCEAMEYLEGNNFVHRDLAARNVLVSEDNVAKVSDFG
LTKEASSTQDTGKLPVKWTAPEALREKKFSTKSDVWSFGILLWEIYSFGRVPYPRIPL
KDVVPRVEKGYKMDAPDGCPPAVYEVMKNCWHLDAAMRPSFLQLREQLEHIKTHELHL
                        "
    intron              161..462
                        /gene="CSK"
                        /number=1
    exon                463..576
                        /gene="CSK"
                        /number=2
    ....
    ....

    polyA_site          5045
                        /gene="CSK"
```

Figure 8.2. The second section of the GenBank flatfile. This is dedicated to the information about the gene and the gene product.

```
BASE COUNT      1084 a    1750 c    1720 g    1099 t
ORIGIN
1 tgcccagtga ggagccagat ctcagcagtt gggggggcatt tcttcacacc cctcctcagt
61 cttcatgctc ttccccacag ctctaatggt accaagtgac aggttggctt tactgtgact
121 cggggacgcc agagctcctg agaagatgtc agcaatacag gtaccacagg ggtgagggtc
181 tgggacatgc aagcattccc accagcccca gcggggtgct tagcagagga gagaggatgc
241 agcttagatc aacccactcc cctttttccc agcactcagt caggtacagg cctgggggaag
301 tgggggagtc tcaacaggaa gggacccagg gctcgttctc cgggcagagc acctcaccca
361 ggctcacaga ggccagctca gaggctgtga ccacgagggt gcgccagcag gtggctggag
421 ggggcccagg ctgcacccgc ccacgtgtca cctgccttgc aggccgcctg gccatccggt
481 acagaatgta ttgccaagta caacttccac ggcactgccg agcaggacct gcccttctgc
541 aaaggagacg tgctcaccat tgtggccgtc accaaggtaa tcaggtgacg cccacccac
601 catcccactg ctgggccttc cctgtgcagg gggaggtggg cctgagagca tgtctgggct
...
...
//
```

Figure 8.2. Continued.

protein ID version increases, but the accession number remains the same. The GI reported in this section refers to the protein translation, and as the gene GI runs parallel to the accession version system, so if the sequence changes a new GI is assigned to the protein translation and the protein ID version will increase. The BASE COUNT reports the numbers of A, G, C and T in the sequence.

8.2 Database search

The aim of a database similarity search is to identify those sequences that are homologous to a query sequence. The basic operation of a search is to align the query to each sequence contained in the database. The results are generally reported in a list of the sequences found, ordered according to a similarity score, plus some statistics.

The most common database searching programs used are the BLAST and FASTA algorithms, available from the NCBI and EBI home pages, respectively.

Table 8.2. A brief description of the programs available in FASTA.

fasta3	Compares protein sequence or DNA sequence to a protein database or to a DNA database respectively.
fastx3/fasty3	Compares a DNA sequence to a protein database, comparing the translated DNA sequence in forward and reverse frames.
tfastx3/tfasty3	Compares a protein to a translated DNA database
fasts3	Compares linked peptides to a protein database
fastf3	Compares mixed peptides to a protein database

8.2.1 FASTA

The FASTA program was the first widely used program for database similarity searching. It applies an improved version (Pearson and Lipman, 1988) of the rapid sequence comparison algorithm described by Pearson and Lipman (Lipman and Pearson, 1985). FASTA is used to compare a protein or DNA sequence to each entry of a database and is available at the EBI server, http://www.ebi.ac.uk/fasta33 (Figure 8.3). A brief description of the available programs is shown in Table 8.2.

Several protein and DNA databases are also available for the search. A complete list is reported at http://www.ebi.ac.uk/fasta33/help.html where details about the algorithm can also be found.

The submission search can be interactive, and the user can read the results on the screen as soon as they are ready, or have the results sent by e-mail.

The user can adjust a few parameters according to his/her needs: the substitution matrix (a series of PAM and BLOSUM matrices is available, the right choice depends on the evolutionary relationship the user wants to find out for his sequence of interest); the ktup value (defining the size of the word to use for the search of the identity regions); the penalties for the introduction and extension of gaps.

Default values are always set but more trials with a different choice of the parameters are recommended.

In this example, the human haemopoietic cell protein-tyrosine kinase gene (HCK, gi number: 183913) is used as the query sequence and the mammalian DNA database is the searched database. The

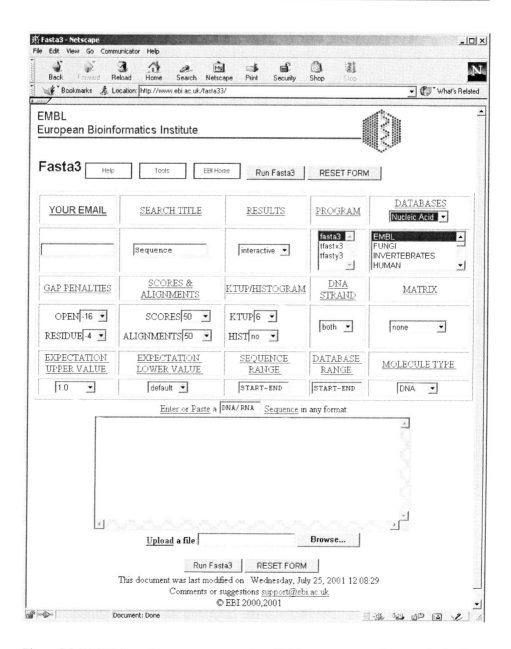

Figure 8.3. FASTA is used to compare a protein or DNA sequence to each entry of a database and is available at the EBI server, http://www.ebi.ac.uk/fasta33.

FASTA output begins with the version of the FASTA program used, the name of the chosen database, the title of the run and the length of the query sequence (Figure 8.4).

```
FASTA searches a protein or DNA sequence data bank
 version 3.3t09 May 18, 2001
Please cite:
W.R. Pearson and D.J. Lipman PNAS (1988) 85:2444-2448

@:1-: 1926 nt
 gi Human hemopoietic cell
 vs  EMBL Mammals library
searching /ebi/services/idata/fastadb/em_om library

40152084 residues in 34274 sequences
  Expectation_n fit: rho(ln(x))= 9.3652+/-0.000285; mu= 4.8190+/-
0.018
 mean_var=166.9001+/-30.462, 0's: 2 Z-trim: 41  B-trim: 0 in 0/85
 Lambda= 0.0993
```

Figure 8.4. FASTA output begins with the version of the FASTA program used, the name of the chosen database, the title of the run and the length of the query sequence.

```
FASTA (3.39 May 2001) function [ optimized, +5/-4 matrix (5:-4)] ktup: 6
 join: 69, opt: 54, gap-pen: -16/ -4, width:  16
 Scan time:  1.940
The best scores are:                              opt bits
E(34310)
EM_OM:MFA320181 AJ320181 Macaca fascicularis m (1515) [ f] 7164 1038       0
EM_OM:HSP320182 AJ320182 Hylobates sp. mRNA fo (1530) [ f] 3445  505 6.4e-
142
EM_OM:SSC277921 AJ277921 Saimiri sciureus part (1527) [ f] 3443  505 7.8e-
142
EM_OM:HSP320183 AJ320183 Hylobates sp. mRNA fo (1605) [ f] 2274  337   2e-91
EM_OM:S81472 S81472 c-yes=proto-oncogene [ dogs (1742) [ f] 1914  286 6.5e-76
EM_OM:AF187884 AF187884 Canis familiaris prote (2947) [ f]  604   98 1.9e-19
EM_OM:FDFMSC J03149 Cat (F.domesticus) c-fms p (3828) [ f]  535   89 1.8e-16
EM_OM:AB066532 AB066532 Macaca fascicularis br (2793) [ f]  482   81 3.4e-14
EM_OM:AF347051 AF347051 Sus scrofa platelet-de ( 781) [ f]  462   78 2.7e-13
EM_OM:S76596 S76596 c-kit=receptor tyrosine ki (4222) [ f]  459   78 3.3e-13
EM_OM:AF044249 AF044249 Canis familiaris recep (3154) [ f]  457   77 4.1e-13
EM_OM:AF099030 AF099030 Canis familiaris KIT ( (2937) [ f]  448   76   1e-12
...
...
```

Figure 8.5. FASTA output. The sequences are ranked according to the evaluated similarity score. For each sequence, the link to the database, the name, the length, the reverse (r) or the forward (f) strand, the *opt* score, the *bit* score and the *e value* are reported.

```
>>EM_OM:HSP320182 AJ320182 Hylobates sp. mRNA for lck pr   (1530
nt)
  initn: 3394 init1: 3394 opt: 3445   Z-score: 2659.7  bits: 505.2
E(): 6.4e-142
  72.883% identity (72.883% ungapped) in 1346 nt overlap (248-
1593:185-1530)

          220        230        240        250        260        270
gi     ACACACCAGGAATCAGGGAGGCAGGCTCTGAGGACATCATCGTGGTTGCCCTGTATGATT
                       : ::::: : : ::  : : :::::: :    :
EM_OM: AAGGCTCCAATCCGCCGGCTTCCCCACTGCAAGACAACCTGGTTATCGCCCTGCACAGCT
           160        170        180        190        200        210

          280        290        300        310        320        330
gi     ACGAGGCCATTCACCACGAAGACCTCAGCTTCCAGAAGGGGGACCAGATGGTGGTCCTAG
       : ::: ::  :::: ::: ::::::  ::::  :::::::::: ::: :    :::: :
EM_OM: ATGAGCCCTCTCACGACGGAGACCTGGGCTTTGAGAAGGGGGAACAGCTCCGCATCCTGG
           220        230        240        250        260        270

...
...

          1540       1550       1560       1570       1580       1590
gi      GTGTGCTGGATGACTTCTACACGGCCACAGAGAGCCAGTACCAACAGCAGCCATGATAGG
       : :::::::: :::::::: :::::::::: ::: :::::::::: :   ::::: :::
EM_OM: GCGTGCTGGAGGACTTCTTCACGGCCACGGAGGGCCAGTACCAGCCTCAGCCTTGA
            1480       1490       1500       1510       1520       1530
```

Figure 8.6. The pair-wise alignments reported by FASTA.

In the next section the sequences are ranked according to the evaluated similarity score. For each sequence, the link to the database, the name, the length, the reverse (r) or the forward (f) strand, the *opt* score, the *bit* score and the *e value* are reported (Figure 8.5). The pair-wise alignments are then reported (Figure 8.6).

It is possible to view the results in a different format. There are three buttons at the top of the output file. The 'View using Mview' button shows the query sequence and the sequences found in the search as a coloured multiple alignment. The first reported sequence is the query sequence, then for each sequence found, the link to the database, the percent identity with the query, and the sequence aligned to the query are reported. The 'Visual Fasta' button shows, before the ranking sequences, a graphical overview of the alignment. The first red line is the query sequence, each sequence found is reported below the query and its colour corresponds to the score obtained in the search.

8.2.2 How FASTA works, a step by step description

The program works in four steps. In the first step, regions of perfect matches in each pair-wise comparison are rapidly searched and listed. In this phase the value of the *ktup* parameter, defining the number of contiguous residues (word) that define a perfect match, is very important for the speed and sensitivity of the search. The default value of *ktup* is 2 for proteins and 6 for DNA sequences, but lower values should be used for finding distant relationships. In the second step of the search, the short identical segments found in each comparison, are re-scored using substitution matrices, and, if necessary, gap scores are introduced. For nucleic acids, positive scores for identity, and negative scores for non-identity are commonly used. In this phase the highest similarity score is calculated and later reported in the output as the *init1* score. In the third step, the initial segments found can eventually be joined together. In this phase only the non-overlapping regions can be joined and the score of the alignment is given by the sum of the scores of each segment minus a penalty for each introduced gap. This score is the *initn* value reported in the output. In the last step, the previous alignment found is optimized using a variation of the Needleman-Wunsch (Needleman and Wunsch,1970) algorithm; the similarity score is then reported in the output as *opt* score. The final alignment reported is obtained applying the Smith-Waterman algorithm (Smith and Waterman, 1981). The statistical significance of each alignment is estimated and reported by two terms, the z-score and the e-value that provide information about the likelihood that the given alignment is occurred purely by chance (Smith *et al.*, 1985).

8.2.3 BLAST

BLAST is available at the NCBI (National Center for Biotechnology Information) site, http://www.ncbi.nlm.nih.gov/BLAST/.

BLAST stands for **B**asic **L**ocal **A**lignment **S**earch **T**ool. The BLAST programs improved the database search speed and put database searching on an accurate statistical foundation (Altschul *et al.*, 1990). The searching unit of the BLAST algorithm is the High-scoring Segment Pair (HSP). In the first step of the search, HSPs between the

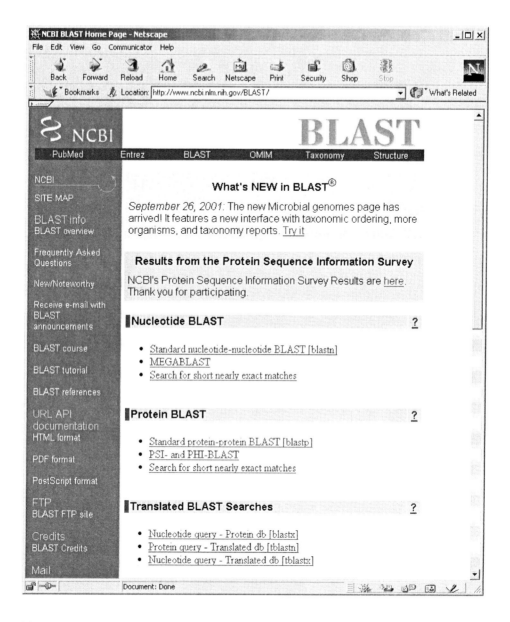

Figure 8.7. The BLAST home page, http://www.ncbi.nlm.nih.gov/BLAST/.

query sequence and each sequence within the database are selected. The BLAST program then evaluates the statistical significance of each match found and finally reports those matches that satisfy a user-selectable threshold of significance. The search of the HSPs segments is achieved by identifying short words, of length determined by the *ktup* value, between the query sequence and each sequence in the database, whose score satisfies a threshold score T. This strategy allows the *ktup* value to remain high for a quick search without loosing sensitivity. Thus, even with a high *ktup* value, it is possible to increase the T score to find closely related sequences of the query sequence. Alternatively, decreasing the T score increases the probability of finding more distantly related sequences. For nucleotide sequences, BLAST considers only perfect matching words, so the sensitivity of the search with respect to distantly related sequences is very low. In such cases the FASTA algorithm is more efficient.

Once a matching word is found, the algorithm tries to extend it in both directions and stops when the similarity score of the segment containing the word decreases significantly.

For each alignment an *e value* and a *bit score* are reported in the output. The *e value* represents, as mentioned before (see the 'Statistical significance of alignments' paragraph chapter 7), the number of alignments with a score equivalent or greater than the reported one expected purely by chance. The *bit score* takes into account the statistical parameters K and lambda of the scoring system and it is normalized with respect to them; therefore, the bit score can be used to compare alignment scores from different searches (Altschul, 1991; Altschul and Gish, 1996; Karlin and Altschul, 1990).

There are several programs based on the BLAST algorithm at the NCBI site. The BLAST home page (Figure 8.7), http://www.ncbi.nlm.nih.gov/BLAST/, briefly describes each program and its use. It is also possible to perform a comparison between two sequences. For nucleotide searches the available programs are listed in Table 8.3.

Table 8.3. BLAST programs for nucleotide searches.

---

blastn compares two nucleotides sequences
tblastn compares a protein sequence against a nucleotide sequence which has been translated in all six frames
blastx compares a nucleotide sequence against a protein sequence
tblastx compares two nucleotides sequences previously translated in all six frames

---

A specialized BLAST section that allows searching of databases related to a specific organism or field of research is available. Sequences can be compared to the human genome, finished and unfinished microbial genomes and the *P. falciparum* genome. VecScreen can be used for identifying vector sequences, and IgBLAST to analyse immunoglobulin sequences in GeneBank.

The sequence of the HCK protein-tyrosine kinase (GI: 183913) human gene was used to search all GenBank+EMBL+DDBJ+PDB sequences (nr) in the example. The BLAST output begins with information about the program used and the reference for BLAST, the title of the query sequence, its length and the chosen database (Figure 8.8).

Next, a graphical overview of the alignment of the sequences that have some similarity with the query sequence is shown (Figure 8.9). The long red line at the top of the graph is the query sequence. Each sequence found is reported below the query sequence, and its colour represents the score obtained. The bar at the top of the graph is

```
BLASTN 2.2.1 [ Apr-13-2001]
Reference:
Altschul, Stephen F., Thomas L. Madden, Alejandro A. Schäffer,
Jinghui Zhang, Zheng Zhang, Webb Miller, and David J. Lipman
(1997),
"Gapped BLAST and PSI-BLAST: a new generation of protein database
search
programs",  Nucleic Acids Res. 25:3389-3402.

RID: 1005130733-125-14962

Query= gi|183913|gb|M16592.1|HUMHCKB Human hemopoietic cell
         (1926 letters)


Database: nt;
            1,000,461 sequences; 70,461,447 total letters


If you have any problems or questions with the results of this search
please refer to the BLAST FAQs
```

Figure 8.8. BLAST output. This begins with information about the program used and the reference for BLAST, the title of the query sequence, its length and the chosen database.

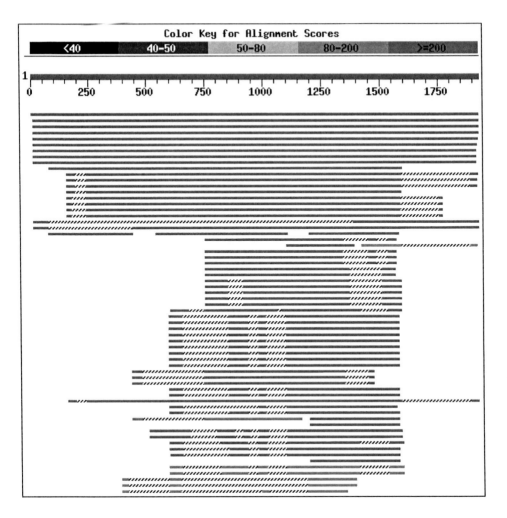

Figure 8.9. A graphical overview of the alignment of the sequences that have some similarity with the query sequence.

a legend and the different colours represent similarity score range. The relative positions of the sequences found and their length are reported. Each one of the sequences aligned in the graph is hyperlinked to its pair-wise alignment with the query.

The next section of the BLAST output file reports the sequences ranked by similarity score (Figure 8.10). For each sequence there is a link to the database, the name of the sequence and the evaluated scores, the *bit score* and the *e value*.

```
Score       E
Sequences producing significant alignments:                          (bits)
Value

gi|183913|gb|M16592.1|HUMHCKB  Human hemopoietic cell protei...  3818  0.0
gi|14772512|ref|XM_009539.4|   Homo sapiens hemopoietic cell ...  3610  0.0
gi|14772508|ref|XM_046236.1|   Homo sapiens hemopoietic cell ...  3610  0.0
gi|14772504|ref|XM_046235.1|   Homo sapiens hemopoietic cell ...  3602  0.0
gi|32045|emb|X58741.1|HSHCKE69  H.sapiens HCK gene for tyros...   361  2e-98
gi|14707730|gb|BC003253.1|BC003253  Mus musculus, clone IMAG...   315  9e-85
gi|32043|emb|X58742.1|HSHCKE11  H.sapiens HCK gene for tyros...   311  1e-83
gi|198942|gb|M57697.1|MUSLYNB  Mouse lyn B protein tyrosine ...   307  2e-82
gi|198940|gb|M57696.1|MUSLYNA  Mouse lyn A protein tyrosine ...   307  2e-82
```

Figure 8.10. BLAST reports the sequences ranked by similarity score.

```
>gi|183913|gb|M16592.1|HUMHCKB Human hemopoietic cell protein-tyrosine
kinase (HCK) gene, complete
                cds, clone HK24
            Length = 1926

 Score = 3818 bits (1926), Expect = 0.0
 Identities = 1926/1926 (100%)
 Strand = Plus / Plus


Query: 1       gaattcctttctaaaatccaaccattccaggaaatagaaatatcaacttggggycttcct
60
               ||||||||||||||||||||||||||||||||||||||||||||||||||||||||||||
Sbjct: 1       gaattcctttctaaaatccaaccattccaggaaatagaaatatcaacttgggggcttcct
60


Query: 61      gagaatgtcagattgatggggtgcatgaagtccaagttcctccaggtcggaggcaataca
120
               ||||||||||||||||||||||||||||||||||||||||||||||||||||||||||||
Sbjct: 61      gagaatgtcagattgatggggtgcatgaagtccaagttcctccaggtcggaggcaataca
120
...

...
Query: 1921 tttacc 1926
            ||||||
Sbjct: 1921 tttacc 1926
```

Figure 8.11. BLAST reports the pair-wise alignments. For each alignment the information about the sequence found in the database, the evaluated e value and bit score, the identities between the sequences and the DNA strand aligned are reported.

The last section of the output reports the pair-wise alignments. For each alignment the information about the sequence found in the database, the evaluated e value and bit score, the identities between the sequences and the DNA strand aligned are reported (Figure 8.11).

8.3 Gene structure prediction

In this section the user will find the description of some of the algorithms used to analyse the structure of the gene in order to identify promoter regions, binding sites of transcription factors, translation initiation sites, the splicing sites and coding regions.

8.3.1 Filters

Eucaryotic genomes generally contain repeated DNA sequences. These elements can have several functions, such as gene regulation, recombination or involvement in the mitotic chromosome movements. These elements can be a source of problems for sequence analysis, especially when a homology search is performed in a nucleotide database such as GenBank.

Two useful programs to mask the repeated elements of DNA sequences are CENSOR (Jurka *et al.*, 1996) and RepeatMasker (Smit, AFA and Green, P., unpublished results). Both algorithms mask the DNA query sequence accessing sets of repeated DNA elements.

8.3.1.1 CENSOR

This program is available at http://charon.girinst.org/Censor_Server-Data_Entry_Forms.html. The user query sequence is compared with a collection of human, rodent and plant repeated elements, and the homologous regions found are then replaced by Xs (Jurka *et al.*, 1996). The algorithm accepts human, rodent, plant and invertebate sequences and the results are sent by e-mail. The entry form page is shown in Figure 8.12, with a box for the user email address and a series of buttons to select the appropriate collection of repeated elements for

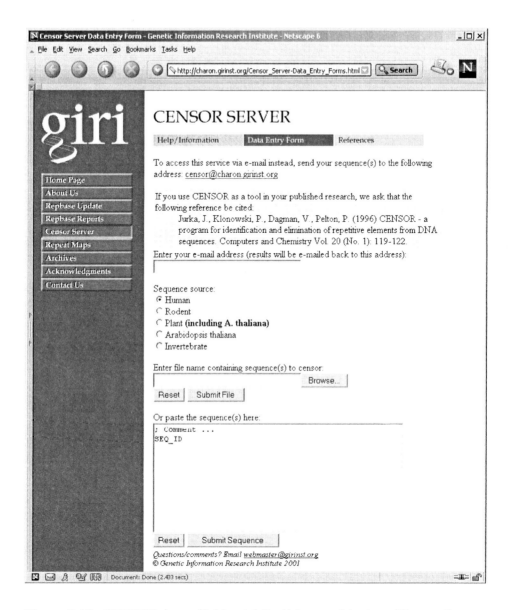

Figure 8.12. CENSOR is available at http://charon.girinst.org/Censor_Server-Data_Entry_Forms.html.

the comparison. Two other boxes allow the user to upload the query sequence or to paste it in the correct format. The query sequence may be in fasta format or in Stanford/IG format (Figure 8.13). A Stanford/

```
;ID    ALU
;DE    Human ALU - a partial consensus
ALU
ggccgggcgcggtggctcacgcctgtaatcccagcacttcaaaaaaaa
```

Figure 8.13. Stanford/IG format suitable for CENSOR input.

IG format consists in one or more lines starting with a semi-colon, reporting a description and/or a comment about the query sequence. A single line follows with the locus name of the sequence (no more than 12 characters and no semi-colon), and then the lines containing the sequence data in lower or upper case.

Output file

The CENSOR server returns the results in a multi-section file displaying the structure shown in Figure 8.14. Here we refer to the output file from the Human toll-like receptor 4 (TLR4) gene (GI: 6175872).

The file starts with some information about CENSOR and a brief description of the overall sections. The query sequence is then reported in order to allow the user to check the submission.

The feature table, *ft.out* (Figure 8.15), follows the query sequence with a summary of the detected repeats.

The value represented by each column (refer to the line in bold) is shown in Table 8.4.

The masked query sequence is then reported in the section 'asap.out', with Xs replacing each base of the repeated region.

```
++——START:name of section ——++
. . .
. . .
. . .
++——END:name of section——++
```

Figure 8.14. The structure of the multi-section results file returned by CENSOR.

The subsequent section of the output file, *local.out* (Figure 8.16), refers to the pair-wise alignments between the query and each repeated elements found in the query sequence. The value of the parameters used to compare the regions and the pair-wise alignments are reported: the '-' symbol indicates an insertion/deletion, '*' indicates identity and '|' identifies a transition and/or a transversion

At the end of each pair-wise alignment, information about the numbers of matches, gaps, transitions and transversions found during the comparison is summarized.

The presence of tandem repeats and/or other repetitive elements is reported in the next section (smpl.out, Figure 8.17).

The subsequent section of the output file, *cmb.out* (Figure 8.18), reports the partial fragments found combined into larger sequence,

```
++——START:  ft.out  ——++
mir              400     519    L1PA12           790     915 d
mir             1416    1950    MER74A             1     556 c
mir             2054    2195    MIR              106     252 c
mir             2350    2460    MIR              137     240 c
mir             2709    3333    L1MB2            235     895 d
mir             3580    3881    MLT1B              7     390 d
mir             4947    5016    PRIMA4_I        7974    8037 c
mir             5697    5742    L2A             2924    2969 d
mir             5747    6055    MER33              1     324 c
mir             6079    6365    L2A             3001    3313 d
mir             6835    7037    L2A             3091    3306 c
mir             7130    7184    MIR3              94     148 d
mir             7279    7350    MIR              180     251 c
mir             7746    7846    L1PA11           816     921 c
mir             8013    8044    (TTTTG)            1      33 d
mir             7942    8109    MIR               31     179 c
mir             8246    8291    (TGGA)             2      47 d
mir             8709    8812    MIR3              81     186 d
mir             9669    9797    MIR3              15     151 d
mir             9932   10228    L1PA16           593     904 c
mir            10857   11036    MIR3              17     207 c
mir            14555   14671    L2B              294     415 c
mir            15833   15875    L1MB1            155     197 d
mir            15879   16750    L1MA10           200    1080 d
mir            17087   17427    MER44A             4     339 c
++——END:  ft.out  ——++
```

Figure 8.15. The CENSOR feature table, *ft.out*, follows the query sequence with a summary of the detected repeats.

```
++——START: local.out ——++
*    Reference Sequence File: sv.humrep.ref.uc
*       Library Sequence File: asap.10250836.11436.subset
*
*    Output file format:
*
*    LOCUS1 N1 N2 LOCUS2 M1 M2 F1 F2 F3 L S #
*
*    ... aligned fragments ...
*
*    ... statistics line from original file ...
*
*    where N1,N2,M1,M2 - aligned fragments boundaries
*      F1 - (no. of Matches)/(no. of Matches + no. of Mismatches + no. of
Gaps),
*      F2 - (Number of Gaps)/(Number of Mismatches),
*           which is set to 0 if (Number of Mismatches) == 0,
*      F3 - (Number of Mismatches)/(Number of Transitions),
*           which is set to 1 if (Number of Transitions) == 0,
*      L  - Length of the top sequence fragment,
*      S  - Local Score.
*
*    Local parameters:
*    Local Algorithm                          =        S-W
*    Margin                                   =        50
*    Min. length to extract insertion         =        12
*    Min. margin to combine fragments         =        30
*    Similarity threshold                     =      22.00
*    Similarity threshold to always keep =         35.00
*    Ratio threshold                          =       2.80
*    Relative Similarity threshold       =          0.23
*    Gap Constant D1                          =       2.95
*    Gap Constant D2                          =       1.90
*    Mismatch Penalty D3                      =      -1.00
*    (Gap penalty = -(D1 + D2 * log(length)))

L1MA10@1      200   900 mir         15879  16577 0.81 0.07   1.57   701
405.34 #
     CCTGTTAGAATGGCTATTATCAAAAAAAACAAAAAATAACAAGTGTTGGCGAGGATGTGGAGAAAWTGGAA
     ** * ************|***         ***** *  ******|| *|*****|*|****** *****
     CCAGGTAGAATGGCTACTAT————AAAAAAATGAAGTGTCATCAAGGATATAGAGAAATTGGAA

     CCCTTGTACACTGTTGGTGGGAATGTAAAATGGTACAGCCACTATGGAAAACAGTATGGAGGTTCCTCAA
     ***** * *****|*** ******* ********||*****||****|********|*******|*** *
     CCCTTCTTCACTGCTGGAGGGAATGGAAAATGGTGTAGCCGTTATGAAAAACAGTACGGAGGTTTCTC-A
     ..
     ..
     TGTTAAGTGTTCT
     *************
     TGTTAAGTGTTCT

and  Containing 564 matches,   9 gaps and 121 mismatches including   77
transitions
++——END: local.out ——++
```

Figure 8.16. The *local.out* section of the output file refers to the pair-wise alignments between the query and each repeated elements found in the query sequence.

Table 8.4. The value represented by each column in the CENSOR feature table.

mir	The name that the user gives to the query sequence
400	Starting position of match in the query sequence
519	Ending position of match in the query sequence
L1PA12	Name of the matching repeat found
790	Starting position of match into the matching repeat found
915	Ending position of match into the matching repeat found
d	Complementary, "c", or direct, "d", strand involved in the match

for example the fragment 15879 to 16750 in *cmb.out* (Figure 8.19) is composed of the fragment from 15879 to 16576 and from 16593 to 16750, separately described in the *plc.out* section. The subsequent sections, cmb.out (Figure 8.18) and plc.out (Figure 8.19) describe the repeated elements found in the query sequence.

The *plc.out* section (Figure 8.19) reports the fragments that were masked with Xs in the query sequence.

The last part of the output file is defined '*annot.out*' (Figure 8.20) and returns the REPBASE reference, that is the database containing the collection of repeated elements used by the CENSOR server and the annotation of all the repeated elements that were found.

```
++——START: smpl.out ——++
;For detailed description of SMPL program, see
;Milosavljevic,A. and Jurka,J. "Discovering simple DNA
;sequences by the Algorithmic significance method,
;Comp. Appl. Biosci., 1993 9(4) 407-411.
;
;output from SMPL run using command:
;   smpl -p 28 -t 20 -w 32 -s 8 seq.10250836.11436 0 0
;
;
++———END: smpl.out ——++
```

Figure 8.17. The presence of tandem repeats and/or other repetitive elements is reported.

```
++——START: cmb.out ——++
;LOCUS          mir
;CM    mir            15879 16576        [15879,16576]            L1MA10@1
[200,899]
;CM    mir            16593 16750        [16593,16750]            L1MA10@1
[922,1080]
;CC    L1MA10 len= 856 sim= 0.00 r= 0.00 Score= 856.00
; REPEAT    mir            15879 ->  16750    L1MA10             200 ->
1080
; FRAGMENT    15879 ->  16750
mir
accaggtagaatggctactataaaaaaatgaagtgtcatcaaggatatagagaaattggaacccttcttc
actgctggagggaatggaaaatggtgtagccgttatgaaaaacagtacggaggtttctcaaaaattaaaa
atagaactgctatatgatccagcaatctcacttctgtatatatacccaaaataattgaaatcagaatttc
aagaaaatatttacactcccatgttcattgtggcactcttcacaatcactgtttccaaagttatggaaac
aacccaaatttccattgaaaaataaatggacaaagaaaatgtgcatatacgtacaatgggatattattca
gcctaaaaaaaggggaatcctgttatttatgacaacatgaataaacccggaggccattatgctatgtaa
aatgagcaagtaacagaaagacaaatactgcctgatttcatttatatgaggttctaaaatagtcaaactc
atagaagcagagaatagaacagtggttcctagggaaaaggaggaagggagaaatgaggaaatagggagtt
gtctaattggtataaaattatagtatgcaagatgaattagctctaaagatcagctgtatagcagagttcg
tataatgaacaatactgtattatgcacttaacattttgttaagagggtacctctcatgttaagtgttctt
accatatacatatacacaaggaagcttttggaggtgatggatatatttattaccttgattgtggtgatgg
tttgacaggtatgtgactatgtctaaactcatcaaattgtatacattaaatatatgcagtttttataatat
caattatgtctgaatgaagctataaaaaagaa1
;LOCUS          mir
;CM    mir            2709 3333          [2709,3333]              L1MB2@1
[235,895]
;CC    L1MB2 len= 625 sim= 0.00 r= 0.00 Score= 625.00
; REPEAT    mir            2709 ->   3333    L1MB2              235 ->
895
; FRAGMENT    2709 ->   3333
mir
aataataagtgttggtgaagatgtgaaaaaatgagaactcctgtacaccatttgtgggaatgtaaaatgg
tacagatgctgtggagaatcatatggtgggtgctcaaaaaattaaaaatagatttaccacatgatccagc
aatctcacttctgagtacgtatccaaaagaattgaaaacagagactttaagagatatttgtacaaccatg
tttatggcagcattattcacaatagctaacgtgtggcaacaatgcaagtgtccatgaacagacaaatgga
taagcaaaatgtggtctatacatacaatggaatattgttcagctttaaaaaggaaggaggctttgatcta
tactacacagaaaagaaccttgaggacattatgcaaagtgaaataagccagtgacaaaaagatacatact
gtatgattccacttctaagagctgcctagagtagtcaagattatagagacaaaagtagtgcatagattca
agggcctagggaaaggggaaatggggagttatttattaatgaatagtggtgatgattgtacaaaaatatg
aacataattaatgccactaaattgtacacatacaaatggtcaagataataaattttatgttatgt1
...
...
++——END: cmb.out ——++
```

Figure 8.18. The section cmb.out reports the partial fragments found combined into larger sequence and describes the repeated elements found in the query sequence.

```
++---START: plc.out ---++
;LOCUS        mir
;ALGNLOCUS    mir              15879 ->  16576    L1MA10@1        200 -
>     899
; FRAGMENT    15879 ->  16576
mir
ACCAGGTAGAATGGCTACTATAAAAAAATGAAGTGTCATCAAGGATATAGAGAAATTGGAACCCTTCTTC
ACTGCTGGAGGGAATGGAAAATGGTGTAGCCGTTATGAAAAACAGTACGGAGGTTTCTCAAAAATTAAAA
ATAGAACTGCTATATGATCCAGCAATCTCACTTCTGTATATATACCCAAAATAATTGAAATCAGAATTTC
AAGAAAATATTTACACTCCCATGTTCATTGTGGCACTCTTCACAATCACTGTTTCCAAAGTTATGGAAAC
AACCCAAATTTCCATTGAAAAATAAATGGACAAAGAAAATGTGCATATACGTACAATGGGATATTATTCA
GCCTAAAAAAAGGGGGAATCCTGTTATTTATGACAACATGAATAAACCCGGAGGCCATTATGCTATGTAA
AATGAGCAAGTAACAGAAAGACAAATACTGCCTGATTTCATTTATATGAGGTTCTAAAATAGTCAAACTC
ATAGAAGCAGAGAATAGAACAGTGGTTCCTAGGGAAAAGGAGGAAGGGAGAAATGAGGAAATAGGGAGTT
GTCTAATTGGTATAAAATTATAGTATGCAAGATGAATTAGCTCTAAAGATCAGCTGTATAGCAGAGTTCG
TATAATGAACAATACTGTATTATGCACTTAACATTTTGTTAAGAGGGTACCTCTCATGTTAAGTGTTC1
;LOCUS        mir
;ALGNLOCUS    mir              16593 ->  16750    L1MA10@1        922 -
>    1080
; FRAGMENT    16593 ->  16750
mir
CACAAGGAAGCTTTTGGAGGTGATGGATATATTTATTACCTTGATTGTGGTGATGGTTTGACAGGTATGT
GACTATGTCTAAACTCATCAAATTGTATACATTAAATATATGCAGTTTTATAATATCAATTATGTCTGAA
TGAAGCTATAAAAAAGAA1
;LOCUS        mir
;ALGNLOCUS    mir               2709 ->   3333    L1MB2@1         235 -
>     895
; FRAGMENT     2709 ->   3333
mir
AATAATAAGTGTTGGTGAAGATGTGAAAAAATGAGAACTCCTGTACACCATTTGTGGGAATGTAAAATGG
TACAGATGCTGTGGAGAATCATATGGTGGGTGCTCAAAAAATTAAAAATAGATTTACCACATGATCCAGC
AATCTCACTTCTGAGTACGTATCCAAAAGAATTGAAAACAGAGACTTTAAGAGATATTTGTACAACCATG
TTTATGGCAGCATTATTCACAATAGCTAACGTGTGGCAACAATGCAAGTGTCCATGAACAGACAAATGGA
TAAGCAAAATGTGGTCTATACATACAATGGAATATTGTTCAGCTTTAAAAAGGAAGGAGGCTTTGATCTA
TACTACACAGAAAAGAACCTTGAGGACATTATGCAAAGTGAAATAAGCCAGTGACAAAAGATACATACT
GTATGATTCCACTTCTAAGAGCTGCCTAGAGTAGTCAAGATTATAGAGACAAAAGTAGTGCATAGATTCA
AGGGCCTAGGGAAAGGGGAAATGGGGAGTTATTTATTAATGAATAGTGGTGATGATTGTACAAAAATATG
AACATAATTAATGCCACTAAATTGTACACATACAAATGGTCAAGATAATAAATTTTATGTTATGT1
...
...
++---END: plc.out ---++
```

Figure 8.19. The section plc.out reports the fragments that were masked with Xs in the query sequence and describes the repeated elements found in the query sequence.

```
++——START: annot.out ——++
;CC * * * * * * * * * * * * * * * * * * * * * * * * * *
;CC                                                    *
;CC    REPBASE UPDATE                                  *
;CC    (C) 1997-2001                                   *
;CC    Genetic Information Research Institute          *
;CC    All rights reserved                             *
;CC                                                    *
;CC    A REFERENCE COLLECTION OF HUMAN REPETITIVE ELEMENTS *
;CC    RELEASE 8.1.4; SEPTEMBER 2001                   *
;CC                                                    *
;CC    Edited by J. Jurka and A.F.A. Smit              *
;CC                                                    *
;CC    OTHER ACTIVE CONTRIBUTORS:                      *
;CC    Kapitonov, V., Walichiewicz, J., Klonowski, P.  *
;CC                                                    *
;CC    CONTRIBUTORS TO EARLIER RELEASES:               *
;CC    Pethiyagoda, C. and Chai A.                     *
;CC                                                    *
;CC    Original release described in:                  *
;CC                                                    *
;CC    Jurka, J., Walichiewicz, J. and Milosavljevic, A. *
;CC    Prototypic Sequences for Human Repetitive DNA   *
;CC    Journal of Molecular Evolution 35:286-291(1992) *
;CC                                                    *
;CC    This release was supported in part by U.S.      *
;CC    Department of Energy grant DE-FG0395ER62139     *
;CC    and by N.I.H. grant 1P41LM06252-01.             *
;CC                                                    *
;CC * * * * * * * * * * * * * * * * * * * * * * * * * *
++———————————————————————————++
;ID    L1MA10        DNA   ; HUM   ; 1080 BP
;XX
;DE    3'-end of L1 repeat (subfamily L1MA10) - a consensus.
;XX
;AC    .
;XX
;DT    01-OCT-1995 (Rel. 3.6, Created)
;DT    07-OCT-1998 (Rel. 6.7, Last updated, Version 3)
;XX
;KW    Repetitive sequence; L1 (LINE) family; L1M3; L1MA10
subfamily.
;XX
;OS    consensus
;XX
```

Figure 8.20. The last part of the CENSOR output file is defined 'annot.out' and returns the REPBASE reference, that is the database containing the collection of repeated elements used by the CENSOR server and the annotation of all the repeated elements that were found.

```
;OC   Homo sapiens
;OC   Eukaryota; Animalia; Metazoa; Chordata; Vertebrata;
Mammalia;
;OC   Theria; Eutheria; Primates; Haplorhini; Catarrhini;
Hominidae.
;XX
;RN   [1]   (bases 1 to 1080)
;RA   Smit,A.F.A., Toth,G., Riggs,A.D. and Jurka,J.
;RT   Ancestral, mammalian-wide subfamilies of LINE-1 repetitive
sequences
;RL   J. Mol. Biol. 246: 401-417 (1995).
;XX
;RN   [2]   (bases 1 to 1080)
;RA   Smit,A.F.A.
;RL   Update (May 3, 2000)
;XX
;CC   ORF2 ends at bp 675; average divergence of copies from
consensus: 16%
;XX
;DR   [2] (consensus)
;XX
;SQ   Sequence 1080 BP; 415 A; 175 C; 214 G; 274 T; 2 other;
++————————————————————————————————++
...
...
++— —END: annot.out ——++
```

Figure 8.20. Continued.

8.3.1.2 ReapeatMasker

The program is available at http://ftp.genome.washington.edu/cgi-bin/ RepeatMasker.

   RepeatMasker screens DNA sequences for repeats as well as for low complexity regions (Smit, AFA and Green, P., unpublished results). It returns an output file with detailed annotation of all the repeats found in the query sequence and the query sequence with the repeats replaced with Ns. The entry form appears as shown in Figure 8.21. The user can submit the query sequence by uploading the file from his/her computer or pasting the sequence in the box in fasta format. The user can also select the format of the output file: if the html button is selected the output file is returned to the user screen; if 'tar' is selected the output file is returned as a UNIX computer

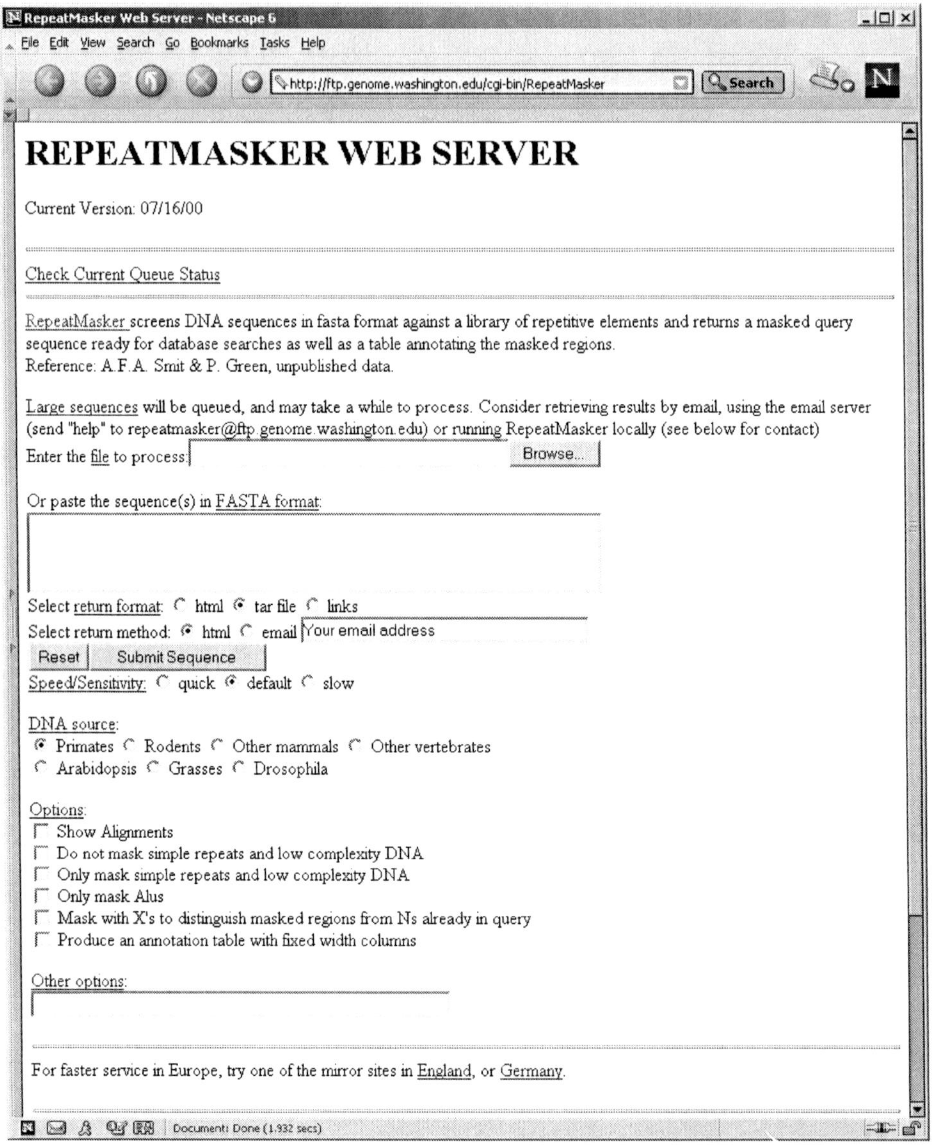

Figure 8.21. RepeatMasker is available at http://ftp.genome.washington.edu/cgi-bin/ RepeatMasker.

compressed file; if 'links' is selected the output files are returned in a text format. The results can be received via email. The speed/sensitivity option may affect the results. The user can choose to run the program at three different levels of speed or sensitivity. The 'slow' setting will take about 3 times longer and will find and mask 0 - 5 % more repetitive DNA sequences than the default setting. The 'quick' setting misses 5 - 10% of the sequences masked by default, but runs 3 to 6 times faster. The DNA source allows the user to indicate the source of his/her query sequence, in order to have a species-specific or a group of species-specific comparisons. The other available options are:

(a) *Show alignments.* This option returns the pair-wise alignments between the query sequence and the repeated elements that were found.

(b) *Do not mask simple repeats and low complexity DNA.* If this option is selected, only interspersed repeats are masked.

(c) *Only mask simple repeats and low complexity DNA.* If selected, only simple repeats and low complexity DNA are masked.

(d) *Only mask Alus.* The user limits the masking and annotation to (primate) Alu repeats.

(e) *Mask with Xs to distinguish masked regions from Ns already in query.* This option allows the user to distinguish the masked regions from the ambiguous bases in the query sequence.

(f) *Produce an annotation table with fixed width columns.* The column width in the annotation table is adjusted to the maximum length of any string occurring in a column; this allows long sequence names to be spelled out completely. If this option is selected, the user can fix the width of the columns.

The 'other options' box of the entry form allows the user to write some less frequently used options in UNIX command style.

Output file

The RepeatMasker server was used to mask the Homo sapiens toll-like receptor 4 (TLR4) gene (GI: 6175872) selecting the html return format and method, the default speed and sensitivity, and the primates and show alignments options. The returned output file consists of four sections, the 'Repeat sequence' (Figure 8.22), the 'Alignments' (Figure 8.23), the 'Masked sequence' and the 'Summary' (Figure 8.24). The 'Repeat sequence' section (Figure 8.22) reports a list of the best matches between the query sequence and any of the sequences in the repeat database or between the query sequence and the low complexity DNA. A match is not shown if 80 % of its domain is contained within a higher scoring match. The matches are ranked by query name, and, for each query, by starting position.

The value represented by each column (see the line in bold) is shown in Table 8.5.

The 'Alignments' section (Figure 8.23) reports the pair-wise alignment between the query and the repeats of the database found within the query sequence.

The first line refers to the same line of the 'Repeat sequence' section, then the alignment is reported, where '-' indicates an insertion/ deletion, 'v' a transversion, 'I' a transition. The last lines refer to other information about the pair-wise alignment, such as the ratio between transitions and transversions.

```
Repeat sequence:

  SW   perc perc perc query       position in query        matching repeat
position in   repeat
score  div. del. ins. sequence begin    end  (left)      repeat    class/
family  begin   end (left)   ID

  673  10.0   5.8  0.8 query       400    519 (18455)  +  L1PA12    LINE/L1
6043 6168      (0)
 2437  21.7   4.5  0.6 query      1416   1950 (17024)  C  MER74A    LTR/ERVL
(2)   556     1
  400  29.3   3.2  0.0 query      2049   2171 (16803)  C  MIR       SINE/MIR
(6)   256     130
  340  27.7   4.0  0.0 query      2350   2450 (16524)  C  MIR       SINE/MIR
(22)   240    136
 2760  17.9   0.9  0.7 query      2701   3243 (15731)  +  L1MB3     LINE/L1
5483 6026  (157)
  443  18.4   0.0  0.0 query      3236   3322 (15652)  +  L1MB4     LINE/L1
6062 6148   (32)     *
. . . . . . . . . .
. . . . . . . . . .
```

Figure 8.22. The 'Repeat sequence' output from RepeatMasker.

Table 8.5. The value represented by each column in RepeatMasker output files.

673	Smith-Waterman score of the match
10.0	% substitutions in the matching region compared to the consensus
5.8	% bases opposite a gap in the query sequence (deleted bp)
0.8	% of bases opposite a gap in the repeat consensus (inserted bp)
Query	name of the query sequence
400	starting position of match in the query sequence
519	ending position of match in the query sequence
(18455)	no. of bases in query sequence past the ending position of match
+	the matching strand
LIPA12	name of the matching repeat
LINE/L1	the class of the repeat
6043	starting position of match in database sequence
6168	ending position of match in database sequence
(0)	number of bases in (complement of) the repeat consensus sequence preceding the beginning of the match (so 0 means that the match extended all the way to the end of the repeat consensus sequence)
*	If present, it indicates that there is a higher-scoring match whose domain partly (<80 %) includes the domain of this match.

```
Alignments:

673   10.00   5.83   0.83   query   400   519   (17143)   L1PA12#LINE/L1   790   915
(0)    4

   query          400 ACTATGCTTAAGAT—GCGATTAATTA—TGTACAACAAACCCCCAT 442
                        v          v i— i    v  v  —
   L1PA12#LINE/L1 790 ACTAGGCTTAATACCTGGGTGATGAAATAATCTGTACAACAAACCCCCAT 839

   query          443 GACACACGTTTACCTATGTAACAAACCTGCTCATCCTGCACATGTACTTC 492
                                                       v              ii
   L1PA12#LINE/L1 840 GACACACGTTTACCTATGTAACAAACCTGCACATCCTGCACATGTACCCC 889

   query          493 TGAATGTAAAAATAAAAGTAAAAAAAA 519
                        iv       —       v
   L1PA12#LINE/L1 890 TGAACTTAAAA-TAAAAGTTAAAAAAA 915

Transitions / transversions = 0.71 (5 / 7)
Gap_init rate = 0.02 (3 / 126), avg. gap size = 2.67 (8 / 3)
```

Figure 8.23. The 'Alignments' output from RepeatMasker.

```
Summary:
==========================================================
file name:  /repeatmasker/tmp/RM2seq
sequences:              1
total length:       18974 bp
GC level:           38.35 %
bases masked:        6562 bp ( 34.58 %)
==========================================================
                number of      length     percentage
                elements*    occupied    of sequence
----------------------------------------------------------
SINEs:               12        1507 bp       7.94 %
        ALUs          0           0 bp       0.00 %
        MIRs         12        1507 bp       7.94 %

LINEs:               12        3263 bp      17.20 %
        LINE1         6        2055 bp      10.83 %
        LINE2         6        1208 bp       6.37 %
        L3/CR1        0           0 bp       0.00 %

LTR elements:         2         837 bp       4.41 %
        MaLRs         1         302 bp       1.59 %
        ERVL          1         535 bp       2.82 %
        ERV_classI    0           0 bp       0.00 %
        ERV_classII   0           0 bp       0.00 %

DNA elements:         2         650 bp       3.43 %
        MER1_type     1         309 bp       1.63 %
        MER2_type     1         341 bp       1.80 %

Unclassified:         0           0 bp       0.00 %

Total interspersed repeats:    6257 bp      32.98 %

Small RNA:            1         106 bp       0.56 %

Satellites:          0           0 bp       0.00 %
Simple repeats:      2          72 bp       0.38 %
Low complexity:      3         127 bp       0.67 %
==========================================================
* most repeats fragmented by insertions or deletions
  have been counted as one element

The sequence(s) were assumed to be of primate origin.
RepeatMasker version 07/16/2000                 default
ProcessRepeats version 07/16/2000
Repbase version 03/31/2000
```

Figure 8.24. The 'Masked sequence' and the 'Summary' output from RepeatMasker.

The 'Masked sequence' section reports the query sequence with the repeats replaced with Ns or Xs according to the user selection. The last section of the output file, the 'Summary' (Figure 8.24), summarizes the information contained in the whole file.

8.3.2 Looking for functional sites in DNA sequences

It is important to realize that actually there is no single tool that can perform all possible analyses on a DNA sequence, and that there are several tools that perform the same analysis. Thus, it is better to submit a sequence to several algorithms in order to compare and combine the results.

There follows a description of some useful algorithms for gene structure prediction.

8.3.2.1 Promoter Scan

The Promoter Scan program was developed for the recognition of polymerase II promoter regions in genomic DNA sequences. It is available at http://bimas.dcrt.nih.gov/molbio/proscan/. The algorithm (Prestridge, 1995) uses a scoring system derived from the analysis of sets of promoter and non-promoter sequences.

The user has only to submit the query sequence in the appropriate box of the Promoter Scan home page (Figure 8.25). There are several sequence formats allowed (http://bimas.dcrt.nih.gov/molbio/readseq/formats.html); the input sequence is first analysed by the Readseq program developed by Don Gilbert (software@bio.indiana.edu) at the Biology Department of Indiana University.

Output file

The Promoter Scan program was used with the promoter region and partial sequence of the human angiotensinogen (AGT) gene (GI: 16417765). The results show the location of the predicted promoter sequences; if the associated scores are higher than a cut-off score set

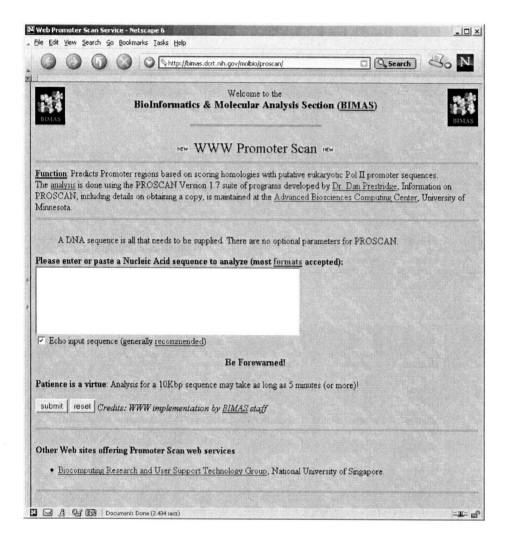

Figure 8.25. The Promoter Scan home page.

to recognize 70 % of primate promoter sequences in the Eukaryotic Promoter Database (Perier *et al.*, 2000). At this cut-off the false positive predictions occur at a rate of approximately one in every 14,000 single strand bases.

The first part of the output (Figure 8.26) reports the query sequence, allowing the user to check the submission. Then the putative promoter sequences are reported:

The first two lines report the sequence range and the strand in

```
Promoter region predicted on forward strand in 957 to 1207
Promoter Score: 53.74 (Promoter Cutoff = 53.000000)
TATA found at 1191, Est.TSS = 1221
Significant Signals:
 Name                      TFD #   Strand  Location  Weight
INF.1                      S01152  +         1122    1.044000
AABS_CS2                   S01612  +         1124    1.012000
AP-1                       S00090  -         1142    1.052000
AP-2                       S00346  +         1177    1.355000
AP-2                       S01936  +         1178    1.108000
PuF                        S02016  -         1185    1.391000
JCV_repeated_sequenc       S01193  -         1185    1.658000
TFIID                      S00087  +         1192    2.618000
TFIID                      S01540  +         1192    1.971000
MLTF                       S00753  +         1206    1.157000
```

Figure 8.26. The first part of the Promoter Scan output.

which the putative promoter is found, then the promoter score and cut-off. If the program finds a TATA box, the TATA box location and the estimated transcription start site (Est.TSS) are reported. The significant signals found follow, most of them are transcriptional elements. The name of the element, the Transcription Factor Database reference number (Ghosh, 2000), the strand, the location and the significance weight of each transcriptional element are reported. The signal weight is based on the ability of that particular signal to discriminate promoter and non-promoter sequences and on the relative frequency with which that signal is found in promoter versus non-promoter sequences.

8.3.2.2 GrailEXP

GrailEXP is a web server that provides several tools developed to predict functional regions in DNA sequences, such as protein coding sequences, promoter regions, and CpG islands. It is available at http://compbio.ornl.gov/grailexp/ for Human, Mouse, Arabidopsis and Drosophila genomes. The microbial genome analysis is provided at http://compbio.ornl.gov/generation/ where the Generation program can work on 37 bacterial genomes. GrailEXP consists of three programs.

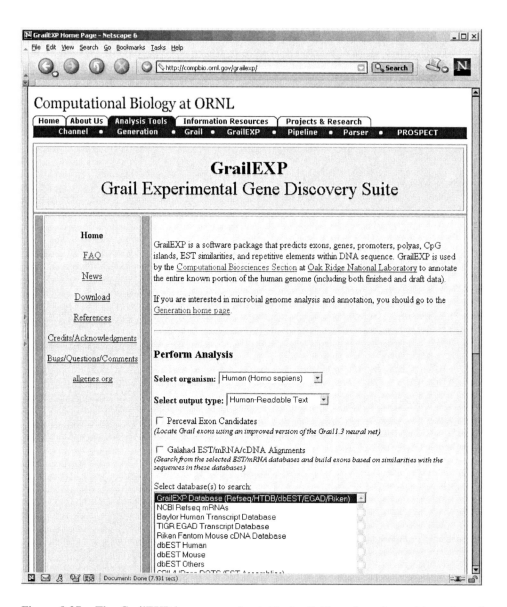

Figure 8.27a. The GrailEXP home page (part A). Available at http://compbio.ornl.gov/ grailexp/.

The first is called Perceval, standing for Protein-coding Exon, Repetitive, and CpG-Island EVALuator, and it is an exon prediction program. The second program is Galahad and it is the gene message alignment program. The gene message alignment consists of an alignment between a genomic sequence, in which genes are loose

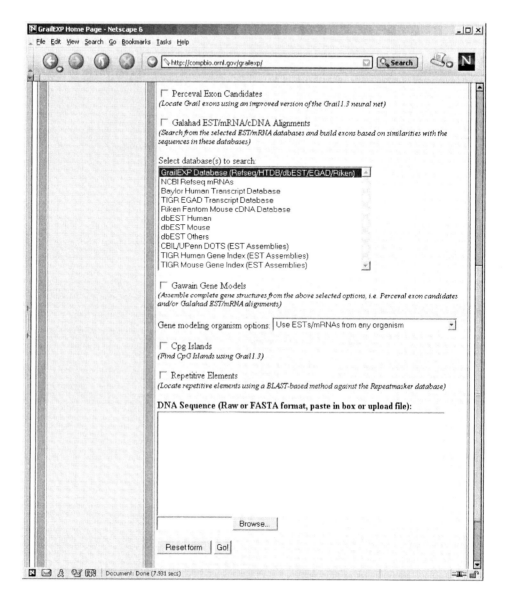

Figure 8.27b. The GrailEXP home page (part B). Available at http://compbio.ornl.gov/grailexp/.

pieces, and a gene message, that is a single piece of DNA built from spliced together exons. The third program is Gawain, standing for Gene Assemblies With Alignment Information. It is the gene assembly program.

It is possible to read a more detailed description of the three programs in the FAQ section at http://compbio.ornl.gov/grailexp/gxpfaq.html.

The home page of GrailEXP is shown in Figure 8.27a and 8.27b. The user first has to select the organism and the output format.

The following choices can affect the results:

Perceval Exon Candidate. If selected, the Perceval program identifies the putative exons and also the CpG islands and the repetitive elements in the query sequence.

Galahad EST/mRNA/cDNA Alignments. If selected, the program searches within the database, chosen by the user among the available ones, in order to build exons according to similarities with sequences contained in the chosen database.

Gawain Gene Models. If selected it uses the information from either Perceval or Galahad or both to model the gene structure.

Cpg Islands and *Repetitive Elements.* If selected, it identifies the CpG islands and the repetitive elements on the query sequence.

The boxes for inserting and uploading the query sequence are reported.

Output file

The Gawain gene predictions results are first displayed. This section reports the information about the gene model. It starts with the FASTA header of the top-scoring reference. Then it continues with the list of all the exons found (Figure 8.28). Lastly, the putative mRNA and coded protein are reported (Figure 8.29).

The following section refers to the Galahad gene message alignment results (Figure 8.30). The alignments reported differ from the classical ones, since they represent boundaries and scores.

For each alignment the 'Seq exons', representing the putative exons of the query sequence, and the 'Ref exons', representing the putative exons of the reference sequence, are shown. Information about the reference sequence is reported at the top of each pair-wise alignment. This alignment consists of a series of number pairs, each one representing the exon/intron boundaries of the relative sequence. In the example the 79..160 pair refers to the exon/intron boundaries of

```
Top-Scoring Reference:  HT1549 (2187 bp) (100% id, 79-5045)
    >human|HT1549|tigr_egad|tyrosine kinase C-SRC, 3'
Reference Path:  HT1549 (2187 bp) (100%, 79-5045)
```

—Index—	——Exons——		——CDS——		-Ph-	-Fr-	-Len-	-Scr-	
1.1.1	79	160	146	160	2	1	82	100
1.1.2	463	576	463	576	0	0	114	100
1.1.3	674	786	674	786	0	1	113	100
1.1.4	1120	1339	1120	1339	2	1	220	100
1.1.5	2260	2353	2260	2353	0	0	94	100
1.1.6	2529	2594	2529	2594	1	1	66	100
1.1.7	2671	2770	2671	2770	1	2	100	100
1.1.8	2860	2950	2860	2950	2	1	91	100
1.1.9	3371	3444	3371	3444	0	1	74	100
1.1.10	3544	3739	3544	3739	2	1	196	100
1.1.11	3846	3932	3846	3932	0	2	87	100
1.1.12	4180	5045	4180	4362	0	0	866	100
PolyA	4994	4999	...	...	.	.	6	87

Figure 8.28. The Gawain gene predictions results. This section reports the information about the gene model. It starts with the FASTA header of the top-scoring reference and continues with the list of all the exons found.

```
>GrailEXP Gene 1, Var 1 mRNA|Similar to HT1549
agctctaatggtaccaagtgacaggttggctttacLgtgactcggggacgccagagctcctgagaagatg
tcagcaatacaggccgcctggccatccggtacagaatgtattgccaagtacaacttccacggcactgccg
agcaggacctgcccttctgcaaaggagacgtgctcaccattgtggccgtcaccaaggaccccaactggta
caaagccaaaaacaaggtgggccgtgagggcatcatcccagccaactacgtccagaagcgggagggcgtg
aaggcgggtaccaaactcagcctcatgccttggttccacggcaagatcacacgggagcaggctgagcggc
ttctgtacccgccggagacaggcctgttcctggtgcgggagagcaccaactaccccggagactacacgct
gtgcgtgagctgcgacggcaaggtggagcactaccgcatcatgtaccatgccagcaagctcagcatcgac
gaggaggtgtactttgagaacctcatgcagctggtggagcactacacctcagacgcagatggactctgta
cgcgcctcattaaaccaaaggtcatggagggcacagtggcggcccaggatgagttctaccgcagcggctg
ggccctgaacatgaaggagctgaagctgctgcagaccatcgggaagggggagttcggagacgtgatgctg
....
....
```

```
>GrailEXP Gene 1, Var 1 protein|Derived from similarity to HT1549
MSAIQAAWPSGTECIAKYNFHGTAEQDLPFCKGDVLTIVAVTKDPNWYKAKNKVGREGIIPANYVQKREG
VKAGTKLSLMPWFHGKITREQAERLLYPPETGLFLVRESTNYPGDYTLCVSCDGKVEHYRIMYHASKLSI
DEEVYFENLMQLVEHYTSDADGLCTRLIKPKVMEGTVAAQDEFYRSGWALNMKELKLLQTIGKGEFGDVM
LGDYRGNKVAVKCIKNDATAQAFLAEASVMTQLRHSNLVQLLGVIVEEKGGLYIVTEYMAKGSLVDYLRS
RGRSVLGGDCLLKFSLDVCEAMEYLEGNNFVHRDLAARNVLVSEDNVAKVSDFGLTKEASSTQDTGKLPV
KWTAPEALREKKFSTKSDVWSFGILLWEIYSFGRVPYPRIPLKDVVPRVEKGYKMDAPDGCPPAVYEVMK
NCWHLDAAMRPSFLQLREQLEHIKTHELHL*
```

Figure 8.29. The putative mRNA and coded protein are reported.

```
GALAHAD Gene Alignments (173 located: 29 displayed, 144 redundant)

 Index Std     Begin        End          Accession      Database
Organism    Length

      1 +          79         1251           BI536292.1    est_others
unknown     483

   4 pieces    Seq exons = (79..160,463..576,674..786,1120..1251)
  90% ident    Ref exons = (43..124,125..238,239..351,352..483)

      2 +          79         2594           BG962606.1     est_mouse
 mouse    882

   6 pieces    Seq exons =
(79..160,463..576,674..786,1120..1351,2273..2353,
                          2529..2594)
  90% ident    Ref exons =
(41..119,120..233,234..346,347..573,574..659,660..730)

      3 +          79         5013           NM_004383.1  refseq_mrna
 human   2420

  12 pieces    Seq exons =
(79..160,463..576,674..786,1120..1339,2260..2353,
                  2529..2594,2671..2770,2860..2950,3371..3444,3544..3739,
                          3846..3932,4180..5013)
  99% ident    Ref exons =
(346..427,428..541,542..654,655..874,875..968,969..1034,
                  1035..1134,1135..1225,1226..1299,1300..1495,1496..1582,
                          1583..2415)
```

Figure 8.30. The Galahad gene message alignment results.

the query sequence and it can be aligned with the 43..124 pair, which refers to the exon/intron boundaries of the reference sequence.

The Perceval results are described in the following section (Figure 8.31). They refer to the exon prediction, the CpG islands and to the repetitive elements.

Each exon is identified by the index number. Then the strand, the start and the end of the putative exon, the frame, the type of the exon, the length, the score and the quality of the prediction are reported.

The Perceval CpG islands identification reports the beginning and the end of the CpG islands, the observed versus expected GC ratio and the percent GC content (Figure 8.32).

```
PERCEVAL Exon Candidates (10 predicted)

Index Std Begin End   Frm   Type  Len   Scr   Quality

     1 +   463   576   0     Internal   114   100   Excellent
     2 +   674   786   0     Internal   113   100   Excellent
     3 +  1120  1339   2     Internal   220   100   Excellent
     4 +  2260  2353   0     Internal    94    84   Good
     5 +  2529  2594   1     Internal    66   100   Excellent
     6 +  2671  2770   1     Internal   100    79   Good
     7 +  2860  2950   2     Internal    91   100   Excellent
     8 +  3371  3444   0     Internal    74    78   Good
     9 +  3544  3739   2     Internal   196    99   Excellent
    10 +  4180  4362   0     Terminal   183    93   Excellent
```

Figure 8.31. The Perceval results.

```
PERCEVAL CpG Islands (1 predicted)

Index     Begin        End         Ratio     Pct_GC

  1       4837        5071         0.80      64.38
```

Figure 8.32. The Perceval CpG islands identification reports the beginning and the end of the CpG islands, the observed versus expected GC ratio and the percent GC content.

```
PERCEVAL Repeats (54 located:  12 simple, 42 complex)

Simple Repeats (12 located)

    Index     Begin        End        Score   1st 10 Bases

      1       1349        1374          64     tctcaggtat...
      2       1526        1555          85     taattttgta...
      3       2636        2674         107     aaaaaaacaa...
      4       5194        5227          85     aaaaaaaaag...
      5       5886        5907         115     aaaaaaaaaa...
      6       8179        8200          61     tttctttctt...
      7      10257       10282         104     tttttttttt...
      8      13734       13775         127     aaaaaaaaaa...
      9      15084       15103          42     taagttaata...
     10      25143       25170          91     tttctttttt...
     11      28313       28346          83     aaaaaaaaaa...
     12      31740       31866         266     aaaaaaaaaa...
```

Figure 8.33. If repetitive elements are found, Perceval reports the complex repeated elements.

```
Complex Repeats (42 located)

Index Std  Begin   End      E-Val   Element Names

  1    +    804     831     4e-08   MIR3#SINE/MIR...
  2    +   1025    1361     8e-232  MSTA#LTR/MaLR/THE1C#LTR/MaLR/T...
  3    +   1400    1490     2e-07   SVA#Other...
  4    +   2392    2863     0e+00   FLAM_C#SINE/Alu/AluJb#SINE/Alu...
  5    +   4918    5193     0e+00   B4A#SINE/B4/PB1D10#SINE/Alu/B1...
  6    +   5617    5885     0e+00   B4A#SINE/B4/PB1D7#SINE/Alu/PB1...
  7    +   7741    7795     6e-11   SVA#Other...
  8    +   8360    8458     9e-04   SVA#Other...
  9    +  10036   10081     1e-05   SVA#Other...
```

Figure 8.34. Using BLAST versus a database of repetitive elements, Perceval finds the complex repeats. For each element found, the strand, the beginning, the end, the e-value of the BLAST search and the element name are then summarized.

If repetitive elements are found, Perceval reports first the simple repeats and then the complex repeated elements (Figure 8.33). For each of the simple repeated elements the beginning, the end, the score and the pattern are reported.

The complex repeats are found using BLAST versus a database of repetitive elements. For each element found, the strand, the beginning, the end, the e-value of the BLAST search and the element name are then summarized (Figure 8.34). This section ends up with the masked query sequence.

8.3.2.3 GenScan

The GenScan software is available at http://genes.mit.edu/ GENSCAN.html (Burge and Karlin, 1997). The server allows the user to identify the location and the exon-intron structure of genes in several organisms. One million base pairs in length is the limit of the accepted sequence length, otherwise a local copy of the program may be requested or the GenScan email server at http://genes.mit.edu/ GENSCANM.html may be used.

The user can submit the query sequence in one-letter code and then has to select the organism and the print options (Figure 8.35).

The 'predicted peptides only' print option returns the translation of the putative coding sequence in the results, while the 'predicted CDS (Coding Sequence) and peptides' option returns the translation and the CDS region of the query sequence.

The results may be returned via email; this is obtained by inserting an email address in the box at the bottom of the GenScan home page (Figure 8.35).

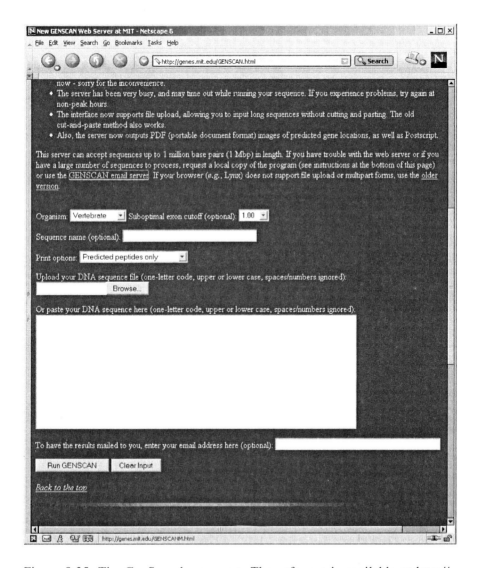

Figure 8.35. The GenScan home page. The software is available at http://genes.mit.edu/GENSCAN.html.

Ouput file

The sequence of the CSK human gene (GI: 402582) was submitted to the server. At the top of the returned output file the information about the GenScan program and the sequence submitted are summarized. The following section refers to the predicted genes/exons found in the query (Figure 8.36).

At the bottom of the output file an explanation about the meaning of each single column is reported (Figure 8.37).

It is possible to download an image of the predicted gene/s in two different formats (postscript and PDF). This figure reports the locations of the exons found in the query sequence (Figure 8.38).

The next section reports the translation of the predicted gene and also the coding sequence (Figure 8.39).

```
Predicted genes/exons:

Gn.Ex Type S .Begin ...End .Len Fr Ph I/Ac Do/T CodRg P.... Tscr..
 —   —   -   —    —   —   —  —   —   —   —   —    —


1.01 Intr +     463    576  114  0  0  107   84   208 0.994  24.11
1.02 Intr +     674    786  113  1  2   76   78   185 0.999  16.65
1.03 Intr +    1120   1339  220  1  1  102   83   565 0.970  56.93
1.04 Intr +    1685   1789  105  1  0   38   80    75 0.904   3.28
1.05 Intr +    2260   2353   94  0  1   84   73   103 0.989   9.10
1.06 Intr +    2529   2594   66  1  0   94   83   138 0.983  13.92
1.07 Intr +    2671   2770  100  2  1   53   86   121 0.999   8.73
1.08 Intr +    2860   2950   91  1  1  101   86   186 0.972  20.86
1.09 Intr +    3371   3444   74  1  2  100   68    35 0.585   2.05
1.10 Intr +    3544   3739  196  1  1   45   81   381 0.999  33.73
1.11 Term +    4180   4362  183  0  0  107   49   344 0.999  30.91
1.12 PlyA +    4994   4999    6                          1.05
```

Figure 8.36. The GenScan output file. This section refers to the predicted genes/exons found in the query.

```
Explanation

Gn.Ex : gene number, exon number (for reference)
Type   : Init = Initial exon (ATG to 5' splice site)
         Intr = Internal exon (3' splice site to 5' splice site)
         Term = Terminal exon (3' splice site to stop codon)
         Sngl = Single-exon gene (ATG to stop)
         Prom = Promoter (TATA box / initation site)
         PlyA = poly-A signal (consensus: AATAAA)
S      : DNA strand (+ = input strand; - = opposite strand)
Begin  : beginning of exon or signal (numbered on input strand)
End    : end point of exon or signal (numbered on input strand)
Len    : length of exon or signal (bp)
Fr     : reading frame (a forward strand codon ending at x has frame
x mod 3)
Ph     : net phase of exon (exon length modulo 3)
I/Ac   : initiation signal or 3' splice site score (tenth bit units)
Do/T   : 5' splice site or termination signal score (tenth bit units)
CodRg  : coding region score (tenth bit units)
P      : probability of exon (sum over all parses containing exon)
Tscr   : exon score (depends on length, I/Ac, Do/T and CodRg scores)
```

Figure 8.37. The GenScan output file. At the bottom of the output file an explanation of the meaning of each single column is reported.

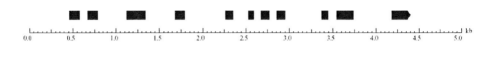

GENSCAN predicted genes in sequence gi

Figure 8.38. The GenScan output. It is possible to download an image of the predicted gene/s in either postscript or PDF format. This image reports the locations of the exons found in the query sequence.

Predicted peptide sequence(s):

Predicted coding sequence(s):

```
>11:10:00|GENSCAN_predicted_peptide_1|451_aa
AAWPSGTECIAKYNFHGTAEQDLPFCKGDVLTIVAVTKDPNWYKAKNKVGREGIIPANYV
QKREGVKAGTKLSLMPWFHGKITREQAERLLYPPETGLFLVRESTNYPGDYTLCVSCDGK
VEHYRIMYHASKLSIDEEVYFENLMQLVEPLTDLHAQKRVQKHEPCALIARAPEALAPAR
TCPQHYTSDADGLCTRLIKPKVMEGTVAAQDEFYRSGWALNMKELKLLQTIGKGEFGDVM
LGDYRGNKVAVKCIKNDATAQAFLAEASVMTQLRHSNLVQLLGVIVEEKGGLYIVTEYMA
KGSLVDYLRSRGRSVLGGDCLLKFSLDVCEAMEYLEGNNFVHRDLAARNVLVSEDNVAKV
SDFGLTKEASSTQDTGKLPVKWTAPEALREKPLKDVVPRVEKGYKMDAPDGCPPAVYEVM
KNCWHLDAAMRPSFLQLREQLEHIKTHELHL
```

```
>11:10:00|GENSCAN_predicted_CDS_1|1356_bp
gccgcctggccatccggtacagaatgtattgccaagtacaacttccacggcactgccgag
caggacctgcccttctgcaaaggagacgtgctcaccattgtggccgtcaccaaggacccc
aactggtacaaagccaaaaacaaggtgggccgtgagggcatcatcccagccaactacgtc
cagaagcgggagggcgtgaaggcgggtaccaaactcagcctcatgccttggttccacggc
aagatcacacgggagcaggctgagcggcttctgtacccgccggagacaggcctgttcctg
gtgcgggagagcaccaactaccccggagactacacgctgtgcgtgagctgcgacggcaag
gtggagcactaccgcatcatgtaccatgccagcaagctcagcatcgacgaggaggtgtac
tttgagaacctcatgcagctggtggagccactcacagacctgcatgctcagaagcgtgtg
cagaaacatgagccatgcgccttgatcgccagggcccctgaggctctcgcacctgcacgg
acatgcccacagcactacacctcagacgcagatggactctgtacgcgcctcattaaacca
aaggtcatggagggcacagtggcggcccaggatgagttctaccgcagcggctgggccctg
aacatgaaggagctgaagctgctgcagaccatcgggaaggggagttcggagacgtgatg
ctgggcgattaccgagggaacaaagtcgccgtcaagtgcattaagaacgacgccactgcc
caggccttcctggctgaagcctcagtcatgacgcaactgcggcatagcaacctggtgcag
ctcctgggcgtgatcgtggaggagaagggcgggctctacatcgtcactgagtacatggcc
aagggagcccttgtggactacctgcggtctagggtcggtcagtgctgggcggagactgt
ctcctcaagttctcgctagatgtctgcgaggccatggaatacctggagggcaacaatttc
gtgcatcgagacctggctgcccgcaatgtgctggtgtctgaggacaacgtggccaaggtc
agcgactttggtctcaccaaggaggcgtccagcacccaggacacgggcaagctgccagtc
aagtggacagcccctgaggccctgagagagaagcccctgaaggacgtcgtccctcgggtg
gagaagggctacaagatggatgcccccgacggctgcccgcccgcagtctatgaagtcatg
aagaactgctggcacctggacgccgccatgcggccctccttcctacagctccgagagcag
cttgagcacatcaaaacccacgagctgcacctgtga
```

Figure 8.39. The GenScan output. This section reports the translation of the predicted gene and also the coding sequence.

8.3.2.4 FGENE

FGENE is available at http://dot.imgen.bcm.tmc.edu:9331/gene-finder/ gf.html and at http://genomic.sanger.ac.uk/gf/gf.shtml (Solovyev *et al.*, 1994; Solovyev and Salamov, 1997). The server offers gene structure prediction for Human, *Drosophila*, *C.elegans*, Yeast, *E.coli* and Plant genomes with different programs.

The query sequence can be pasted in plain text or fasta format. The user has to select the organism and a search method (Figure 8.40). The programs available are shown in Table 8.6.

Output file

The program FGENE was tested with the human gene for the CSK protein kinase (GI: 402582). The output file return is shown in Figure 8.41.

The first lines refer to the query sequence, reporting its first three lines. Then the output file summarizes the name and length of the query sequence and the number of putative exons found. The file reports for each exon a line as shown in Figure 8.42.

The output file ends with the coding sequence. It displays the length in base pairs and residues and the translation.

Table 8.6. The programs available in FGENE.

FGENES	Based on pattern recognition of different types of exons, promoters and polyA signals for the prediction of multiple genes in Human DNA sequences
FGENESH	Algorithm based on Hidden Markov Models for the prediction of multiple genes in Human DNA sequences
FGENES-M	Similar to FGENES, but it predicts several suboptimal variants of the predicted gene structure
BESTORF	Predicts potential coding fragment in EST/mRNA sequence in Human, Drosophila and Plant DNA sequences
FGENE	Predicts gene structure in Human DNA sequences
FEX	Predicts internal, 5'- and 3'- exons in Human DNA sequences
SPL	Predicts splice sites in Human DNA sequences
CDSB	Predicts the protein coding regions in E.coli sequences
NSITE	Searches for regulatory regions
POLYAH	Recognizes the 3'-end cleavage and polyadenilation region of human mRNA precursors
TSSG	Predicts the human PolII promoter regions and the transcription start sites
TSSW	Predicts the human PolII promoter regions and the transcription start sites
RNASPL	Predicts the exon-exon junction positions in cDNA sequences
HBR	Recognizes the Human and E.coli sequences to test a library for *E. coli* contamination

---

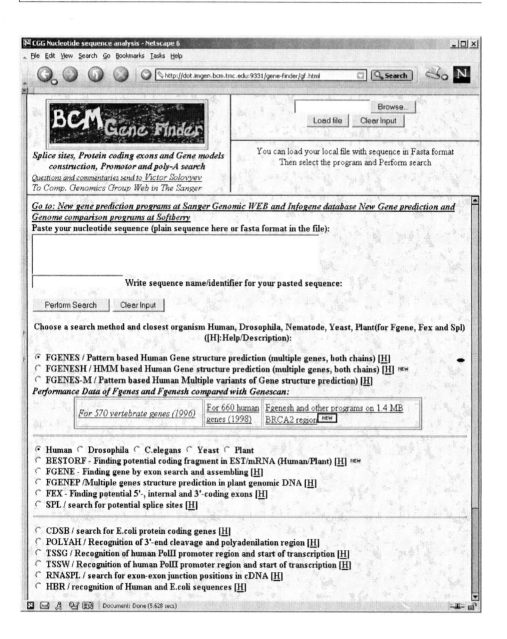

Figure 8.40. FGENE is available at http://dot.imgen.bcm.tmc.edu:9331/gene-finder/gf.html and at http://genomic.sanger.ac.uk/gf/gf.shtml.

```
Name: my_sequence
First three lines of sequence:
TGCCCAGTGAGGAGCCAGATCTCAGCAGTTGGGGGGCATTTCTTCACACCCCTCCTCAGTCTTCATGCTCTTCCC
CACAGCTCTAATGGTACCAAGTGACAGGTTGGCTTTACTGTGACTCGGGGACGCCAGAGCTCCTGAGAAGATGTC
AGCAATACAGGTACCACAGGGGTGAGGGTCTGGGACATGCAAGCATTCCCACCAGCCCCAGCGGGGTGCTTAGCA
fgene  Wed Oct 31 12:28:59 GMT 2001

>my_sequence
 length of sequence -    5653
 number of predicted exons - 12
 positions of predicted exons:
     146 -      160 w=    2.30  ORF:    146 -        160
     463 -      576 w=   13.52  ORF:    463 -        576
     674 -      786 w=   14.45  ORF:    674 -        784
    1120 -     1339 w=   14.75  ORF:   1121 -       1339
    1685 -     1789 w=    5.11  ORF:   1685 -       1789
    2260 -     2353 w=    5.86  ORF:   2260 -       2352
    2529 -     2594 w=   16.08  ORF:   2531 -       2593
    2671 -     2770 w=    8.43  ORF:   2673 -       2768
    2860 -     2950 w=   11.04  ORF:   2861 -       2950
    3371 -     3444 w=    4.03  ORF:   3371 -       3442
    3544 -     3739 w=   10.45  ORF:   3545 -       3739
    4180 -     4362 w=    9.81  ORF:   4180 -       4362
 Length of Coding region-    1371bp  Amino acid sequence -       456aa
MSAIQAAWPSGTECIAKYNFHGTAEQDLPFCKGDVLTIVAVTKDPNWYKAKNKVGREGII
PANYVQKREGVKAGTKLSLMPWFHGKITREQAERLLYPPETGLFLVRESTNYPGDYTLCV
SCDGKVEHYRIMYHASKLSIDEEVYFENLMQLVEPLTDLHAQKRVQKHEPCALIARAPEA
LAPARTCPQHYTSDADGLCTRLIKPKVMEGTVAAQDEFYRSGWALNMKELKLLQTIGKGE
FGDVMLGDYRGNKVAVKCIKNDATAQAFLAEASVMTQLRHSNLVQLLGVIVEEKGGLYIV
TEYMAKGSLVDYLRSRGRSVLGGDCLLKFSLDVCEAMEYLEGNNFVHRDLAARNVLVSED
NVAKVSDFGLTKEASSTQDTGKLPVKWTAPEALREKPLKDVVPRVEKGYKMDAPDGCPPA
VYEVMKNCWHLDAAMRPSFLQLREQLEHIKTHELHL*
```

Figure 8.41. The FGENE output file.

```
     674 -      786 w=   14.45  ORF:    674 -        784
```

```
 674    : starting nucleotide of the exon.
 786    : ending position of the exon.
 14.45  : weight associated to the exon.
 674    : number of the first codon of the exon.
 784    : number of the last codon of the exon.
```

Figure 8.42. The FGENE output file reports for each exon a line as shown.

8.3.2.5 GeneMark

GeneMark is available at http://opal.biology.gatech.edu/GeneMark/, this is the home page of a family of gene prediction programs, listed in Table 8.7.

The sequence of the human gene for the CSK tyrosine kinase (GI:402582) was used with the GeneMark.hmm program. The submission form appears as in Figure 8.43. The title used for the results is optional, the input sequence can be pasted in Genbank, FASTA, GCG, EMBL and raw data formats or can be uploaded by using the 'Sequence File upload' box. Then it is useful to select the species of the query sequence; the organisms available are: *H. sapiens, C. elegans, D. melanogaster, A. thaliana, C. reinardtii, G. gallus* and *O.sativa.*

It is also possible to suggest extension of the prediction to an organism by sending an email to john@amber.biology.gatech.edu.

GeneMark.hmm requires an email address if the user wants the graphical output of the results or if he/she submits a sequence longer than 1,000,000 base pairs.

At this step the user can choose to:

- Select the checkbox to receive by email the graphical view of the results.
- Select the checkbox to run GeneMark after running GeneMark.hmm.
- Select the checkbox to translate the predicted coding region/s found in fasta format.

Output file

The first lines report information about the program used and the sequence submitted (Figure 8.44). The information about the putative exons found is then reported (Figure 8.45).

This section shows the numbers of exons found, then the strand, the type, the starting and ending positions, the length and the frame of each exon.

Table 8.7. The GeneMark family of gene prediction programs.

GeneMark	http://opal.biology.gatech.edu/GeneMark/genemark24.cgi (Borodovsky and McIninch, 1993)
GeneMark.hmm	http://opal.biology.gatech.edu/GeneMark/gmhmm2_prok.cgi for prokaryotes and simple eukaryotes (Lukashin and Borodovsky, 1998) http://opal.biology.gatech.edu/GeneMark/eukhmm.cgi for eukaryotes (Borodovsky and Lukashin, unpublished)
Heuristic GeneMark.hmm	http://opal.biology.gatech.edu/GeneMark/heuristic_hmm2.cgi (Besemer and Borodovsky, 1999)
GeneMark.fbf	http://opal.biology.gatech.edu/GeneMark/fbf.cgi (Shmatkov et al., 1999)
GeneMarkS	http://opal.biology.gatech.edu/GeneMark/genemarks.cgi (Besemer et al., 2001)
GeneMarkS EV	http://opal.biology.gatech.edu/GeneMark/genemarks_ev.cgi (Besemer et al., 2001)

If the checkbox for the protein translation is selected, the next section of the file reports the putative protein in fasta format (Figure 8.46).

If the user chooses to run GeneMark after GeneMark.hmm, the results are then reported in the file (Figure 8.47).

The first lines refer to the program and to query sequence. Then the list of the regions of interest, representing the region between two stop codons in the same reading frame with a coding potential region, is reported (Figure 8.48). The last section of the output file shows the list of the putative exons (Figure 8.49).

If the checkbox for graphical output is selected, the GeneMark server sends an email to the user with a postscript file. This file is a graph reporting the location of the putative exons found. Both the direct and the complementary strand are reported, each one in three reading frames (see an example in Figure 8.50).

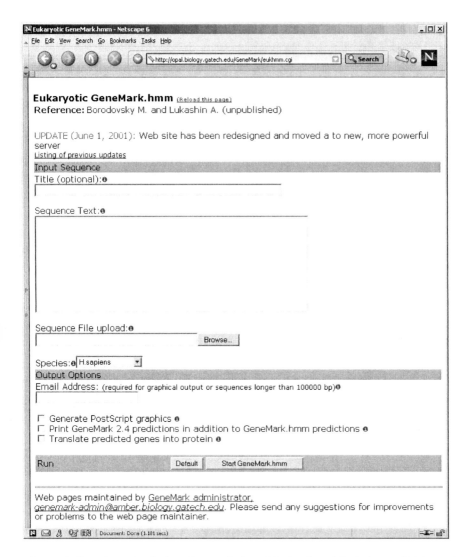

Figure 8.43. The submission form of GeneMark. GeneMark is available at http://opal.biology.gatech.edu/GeneMark/.

```
GeneMark.hmm (Version 2.2a)
Sequence name: my sequence
Sequence length: 5653 bp
G+C content: 61.38%
Matrix: Homo sapiens
Thu Nov  1 05:12:56 2001
```

Figure 8.44. The GeneMark output file. The first lines report information about the program used and the sequence submitted.

Gene #	Exon #	Strand	Exon Type	Exon Range		Exon Length	Start/End Frame
1	1	+	Internal	463	576	114	1 3
1	2	+	Internal	674	786	113	1 2
1	3	+	Internal	1120	1339	220	3 3
1	4	+	Internal	2260	2353	94	1 1
1	5	+	Internal	2529	2594	66	2 1
1	6	+	Internal	2671	2770	100	2 2
1	7	+	Internal	2860	2950	91	3 3
1	8	+	Internal	3371	3444	74	1 2
1	9	+	Internal	3544	3739	196	3 3
1	10	+	Terminal	4180	4362	183	1 3

Figure 8.45. The GeneMark output file. Information about the putative exons.

```
>my sequence_1|GeneMark.hmm|gene 1|416_aa
AAWPSGTECIAKYNFHGTAEQDLPFCKGDVLTIVAVTKDPNWYKAKNKVGREGIIPANYV
QKREGVKAGTKLSLMPWFHGKITREQAERLLYPPETGLFLVRESTNYPGDYTLCVSCDGK
VEHYRIMYHASKLSIDEEVYFENLMQLVEHYTSDADGLCTRLIKPKVMEGTVAAQDEFYR
SGWALNMKELKLLQTIGKGEFGDVMLGDYRGNKVAVKCIKNDATAQAFLAEASVMTQLRH
SNLVQLLGVIVEEKGGLYIVTEYMAKGSLVDYLRSRGRSVLGGDCLLKFSLDVCEAMEYL
EGNNFVHRDLAARNVLVSEDNVAKVSDFGLTKEASSTQDTGKLPVKWTAPEALREKPLKD
VVPRVEKGYKMDAPDGCPPAVYEVMKNCWHLDAAMRPSFLQLREQLEHIKTHELHL
```

Figure 8.46. The GeneMark output file. If the checkbox for the protein translation was selected the file reports the putative protein in fasta format.

```
Sequence: my sequence
Sequence file: gm_sequence
Sequence length: 5653
GC Content:  61.38%
Window length: 96
Window step: 12
Threshold value: 0.500
Matrix: H. sapiens, 0.00 < GC < 0.46 - Order 4
Matrix author: JDM
Matrix order: 4
```

Figure 8.47. The GeneMark output file. If the user chooses to run GeneMark after GeneMark.hmm, the results are then reported in the file.

```
List of Regions of interest
(regions from stop to stop codon w/ a signal in between)
    LEnd        REnd      Strand         Frame

        22         222  complement     fr 3
      1094        1438  direct         fr 2
      3446        3859  direct         fr 2
      4123        4362  direct         fr 1
```

Figure 8.48. The GeneMark output file. The list of the regions of interest, representing the region between two stop codons in the same reading frame with a potential coding region.

```
List of Protein-Coding Exons
(regions between acceptor and donor site w/ coding function
>0.500000)
   Left       Right
   End        End        Strand       Frame    Prob

       73        150  complement     fr 3    0.6244
       84        126                         0.7289

      485        549  direct         fr 1    0.7637

     1221       1333  direct         fr 2    0.8480
     1240       1314                         0.9493

     2671       2719  direct         fr 3    0.5793

     3557       3703  direct         fr 2    0.7925
     3575       3672                         0.8709

     4222       4328  direct         fr 1    0.5712
```

Figure 8.49. The GeneMark output file. The list of the putative exons.

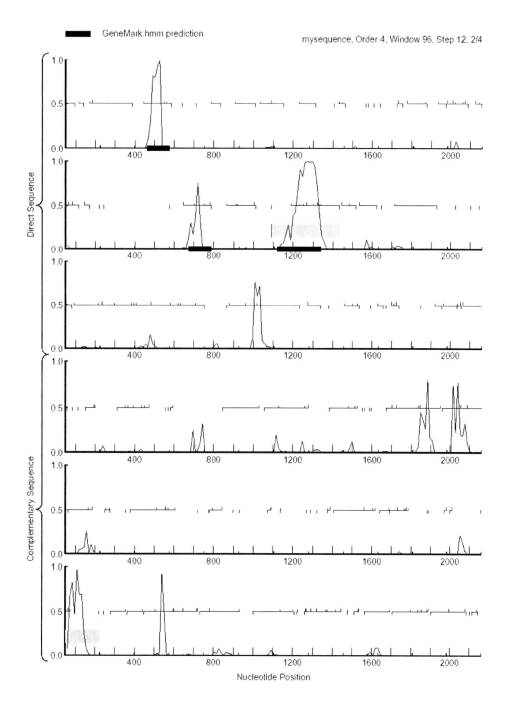

Figure 8.50. The GeneMark output file. If the checkbox for graphical output is selected, the GeneMark server sends an email to the user with a postscript file. This file is a graph reporting the location of the putative exons found. Both the direct and the complementary strand are reported, each one in three reading frames.

8.3.2.6 WebGene

WebGene offers several tools for the analysis and prediction of gene structure. The URL of this site is http://www.itba.mi.cnr.it/webgene/. The tools available are listed in Table 8.8.

GeneBuilder has been used with the sequence of the human gene CSK (GI:402582). This program (Milanesi and Rogozin, 1998; Milanesi *et al.*, 1999) is composed of several modules and predicts gene structure using different approaches together with similarity searches in proteins and EST databases. The home page appears as in Figure 8.51a and 8.51b. The user can select several parameters/options to refine the gene structure prediction. GeneBuilder is available for Human, Mouse, *Fugu, Drosophila, C. elegans, Arabidopsis* and *Aspergillus* genomes.

Mode. This option allows the user to have the full gene model or just the exons with the best scores.

Strand. For choosing the strand on which the prediction is performed.

Sequencing error correction. This option allows the user to handle potential sequencing errors due to frame-shifts and substitutions in the stop codons. The prediction is improved if these errors are eliminated selecting the automatic correction option.

Splice sites prediction. There are two options, 'all' and 'excellent only'. Selecting the 'all' option the program finds 98 % of the real splicing sites but 30 - 35% of all sites are false positives, selecting the 'excellent only' option the prediction goes to 95 % with 15 % false positives.

Potential coding regions. With the 'all' option selected the gene structure prediction is made using all the putative coding exons. With the 'good' option only the exons with good scores are used for the gene model. With the 'excellent' option only the 'excellent' exons are chosen and with the 'key protein similarity' option the exons used are the ones with similarity to a selected homologous protein.

First and last coding exons. This option allows the determination of

Table 8.8. WebGene tools for the analysis and prediction of gene structure.

GeneBuilder	A program for the analysis and prediction of gene structure. (Milanesi and Rogozin, 1998; Milanesi *et al.*, 1999)
ORFGene	A program for the gene structure prediction using information on homologous protein sequence (Rogozin *et al.*, 1996; Milanesi and Rogozin, 1998; Milanesi *et al.*, 1999)
ESTMap	An EST mapping program. (Milanesi and Rogozin, 1998; Milanesi *et al.*, 1999)
Syncod	A program for coding sequence prediction based on BLASTN output. (Rogozin *et al.*, 1999)
RepeatView	A program for the identification of repeated elements. (Milanesi and Rogozin, 1998; Milanesi *et al.*, 1999)
CpG	Identifies the CpG islands. (Milanesi and Rogozin, 1998; Milanesi *et al.*, 1999)
SpliceView	Offers splice site identification. (Rogozin and Milanesi, 1997)
HCpolya	A program for poly-A signals prediction in 3' regions of eukaryotic sequences. (Milanesi *et al.*, 1995; Milanesi *et al.*, 1996; Milanesi and Rogozin, 1998; Milanesi *et al.*, 1999)
HCtata	A program for TATA box identification in eukaryotic sequences. (Milanesi *et al.*, 1995; Milanesi *et al.*, 1996; Milanesi and Rogozin, 1998; Milanesi *et al.*, 1999)
GenView	A program for protein-coding gene prediction (Milanesi *et al.*, 1993a, 1993b; Rogozin *et al.*, 1995)
AUG	A program for start codon prediction. (Rogozin *et al.*, 2001)

the first and last exon of a putative gene using similarity to a chosen protein.

Sequence segment for coding regions prediction. The user can insert the range of the sequence region where the program estimates the presence of the exons. This is useful when long sequences are submitted.

Complete gene model. The 'yes' option returns only the complete potential gene structure, while the 'no' option returns the model of any gene found, including the partial genes.

Use repeated elements mapping. The 'yes' option allows the program to search and mask the repeated elements.

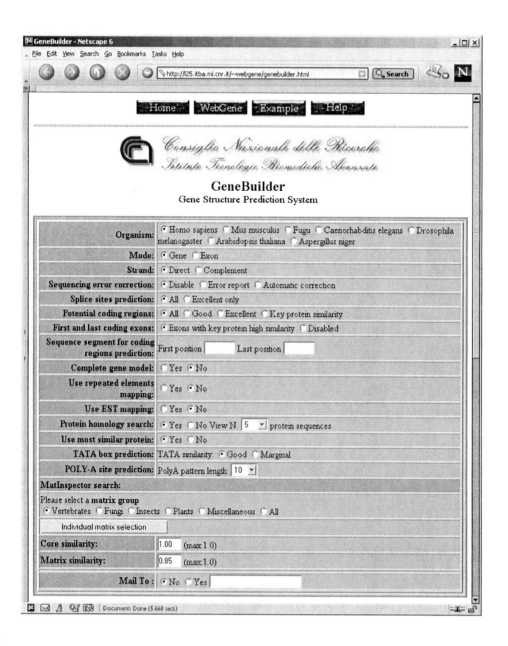

Figure 8.51a. The GeneBuilder home page (part A).

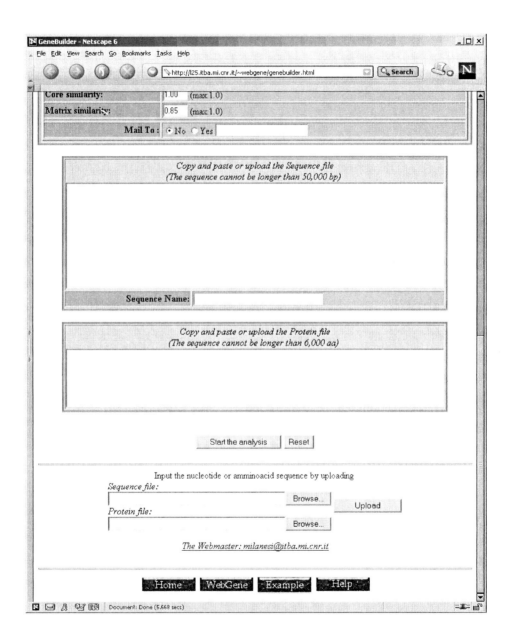

Figure 8.51b. The GeneBuilder home page (part B).

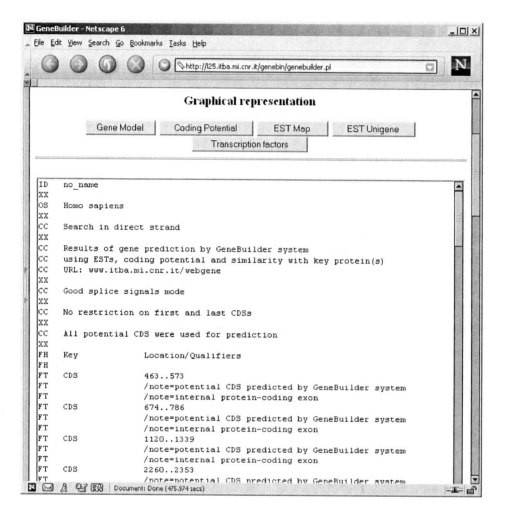

Figure 8.52. The GeneBuilder output file. This section displays the coding regions, the EST matching regions, the translated peptide and the query sequence.

Use EST mapping. The 'yes' option allows GeneBuilder to perform a similarity search versus the EST database.

Protein homology search. This option and the next one allow protein database searching.

TATA box prediction. This option allows the TATA-box prediction. The 'good' option returns only the best putative TATA-box, while the

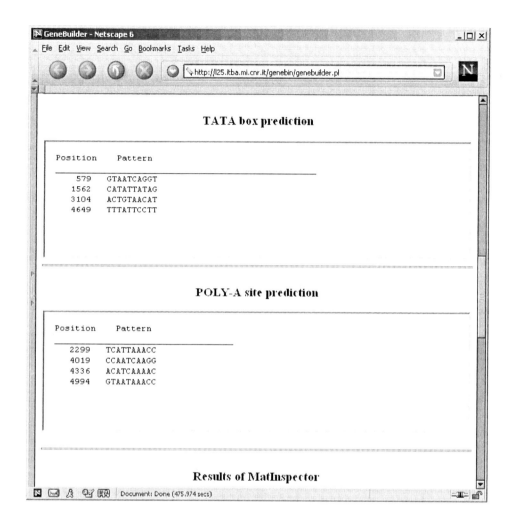

Figure 8.53. The GeneBuilder output file. This section displays the position and the pattern of the TATA-box and the poly-A signals.

'marginal' option returns all the potential TATA-boxes.

POLY-A site prediction. This option allows poly-A signals prediction. The box for the length of this signal can be used to increase or to decrease the pattern discrimination.

MatInspector search. With this option potential transcription factor sites may be identified.

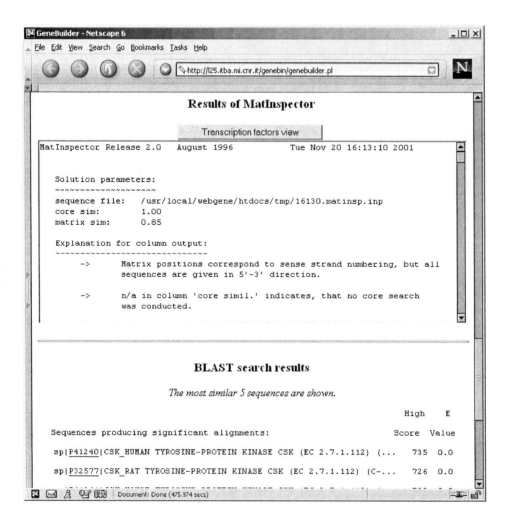

Figure 8.54. The GeneBuilder output file. This section displays the results of MatInspector, and the potential transcription factor sites found are then reported.

Output file

The returned output file is composed of several windows reporting the results of the GeneBuilder modules (Figure 8.52, Figure 8.53, Figure 8.54 and Figure 8.55). In the first output box (Figure 8.52), the coding regions, the EST matching regions, the translated peptide and the query sequence are displayed. In the following two boxes, the position and the pattern of the TATA-box and the poly-A signals are

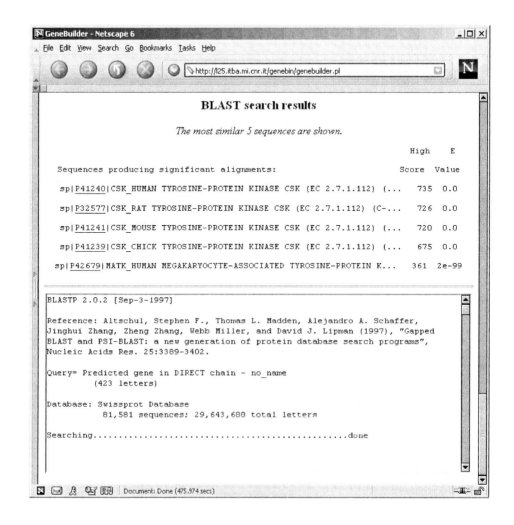

Figure 8.55. The GeneBuilder output file. This section displays the the SWISSPROT search. The most similar 5 sequences can be retrieved from the SWISSPROT database.

shown (Figure 8.53). The next box refers to the results of MatInspector, and the potential transcription factor sites found are then reported (Figure 8.54). The last section of the output file refers to the SWISSPROT search (Figure 8.55). The results of a BLASTP search are also summarized. The most similar 5 sequences can be retrieved from the SWISSPROT database.

It is possible to view a graphical representation of the results clicking on the buttons at the top of the file (Figure 8.52).

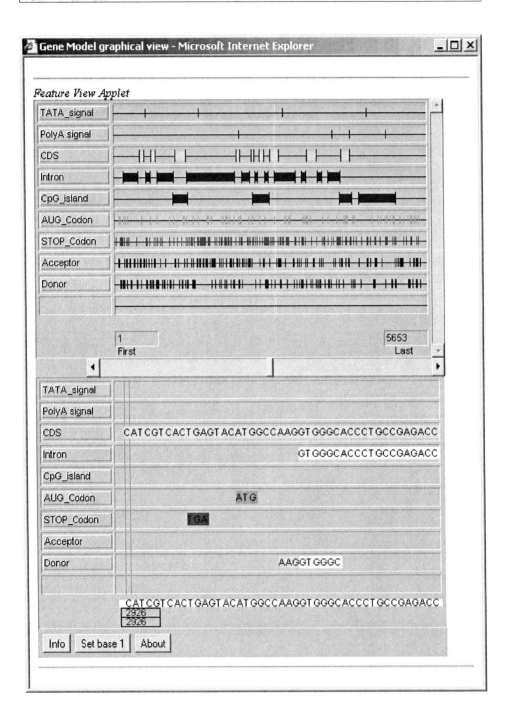

Figure 8.56. The GeneBuilder output file. An example of the graphical representation. The figure at the top reports the functional sites over all the query sequence, while the figure at the bottom gives a more detailed view of the sequence and the sites found.

Gene Model. Gives an overview of the query sequence and all the functional sites found.

Coding Potential. Gives an overview only of the coding regions found.

EST Map. It gives an overview of the EST regions found.

Transcription Factors. It gives an overview of the transcription factor binding sites found.

Each graphical representation consists of two figures. The figure at the top reports the functional sites over all the query sequence, while the figure at the bottom gives a more detailed view of the sequence and the sites found (see an example in Figure 8.56).

8.3.2.7 GeneId

GeneId is available at http://www1.imim.es/geneid.html (Guigo *et al.*, 1992). GeneId is a program for gene structure prediction. First of all, splicing sites, start and stop codons are identified, then the exons are defined and finally the structure of the gene/genes is predicted.

GeneId can be divided into three sections: the 'Input data', the 'Prediction options' and the 'Output options'. The input data section consists of the box for inserting the query sequence in fasta format, the box to upload the sequence from the local user computer and a checkbox that allows the user to grasp a graph of the results in PostScript format.

The Prediction options section consists of two options, one for the organism, Human or Drosophila, and one to select among (1) gene prediction, allowing only the prediction of exons and selected signals, (2) reverse-strand predictions and (3) forward-strand predictions.

The last section is the Output option that allows the user to select the output format and the elements to include in the output file, such as the acceptor and donor sites.

8.3.2.8 PROCRUSTES

PROCRUSTES is available at http://www-hto.usc.edu/software/procrustes/wwwserv.html.

The home page, which contains a description of the software and a manual on how to use the server, is available at http://www-hto.usc.edu/software/procrustes/index.html. Procrustes is a gene prediction software based on the spliced alignment algorithm. This algorithm works on a query sequence and on a protein related to it. It explores all the exon candidates trying to find the best fit with the related target protein (Gelfand 1996). For each query sequence it is possible to use up to 10 different related target proteins and for each query-sequence/target-protein pair Procrustes makes a gene prediction.

The home page of the WWW server appears as shown in Figure 8.57a and 8.57b. It is possible to run the program in a basic and in a test mode, to verify the quality of the prediction.

The submission form consists of boxes where the user can insert the query sequence and the target related sequences (up to 10 sequences). At the top there is also a box for email address if the user wants to receive the results via email.

At the bottom of the page, the parameters, that Procrustes utilizes to perform the alignment between the query and the related sequences, can be changed by the user. These parameters are: the gap open penalty and the gap extension penalty (see Chapter 7), the alignment mode that allows the user to choose between local and global alignment (see Chapter 7). The substitution matrix used to compare the protein sequences is PAM120; at the moment it is the only available matrix. The user can influence the choice of the candidate exons working on the filtration parameter. Procrustes starts the sequence analysis by building a set of candidate exons that then have to pass a filter. This filter consists of a statistical analysis of codon usage and splicing sites. The weak filtration is recommended for close sequences and the moderate one for distant targets.

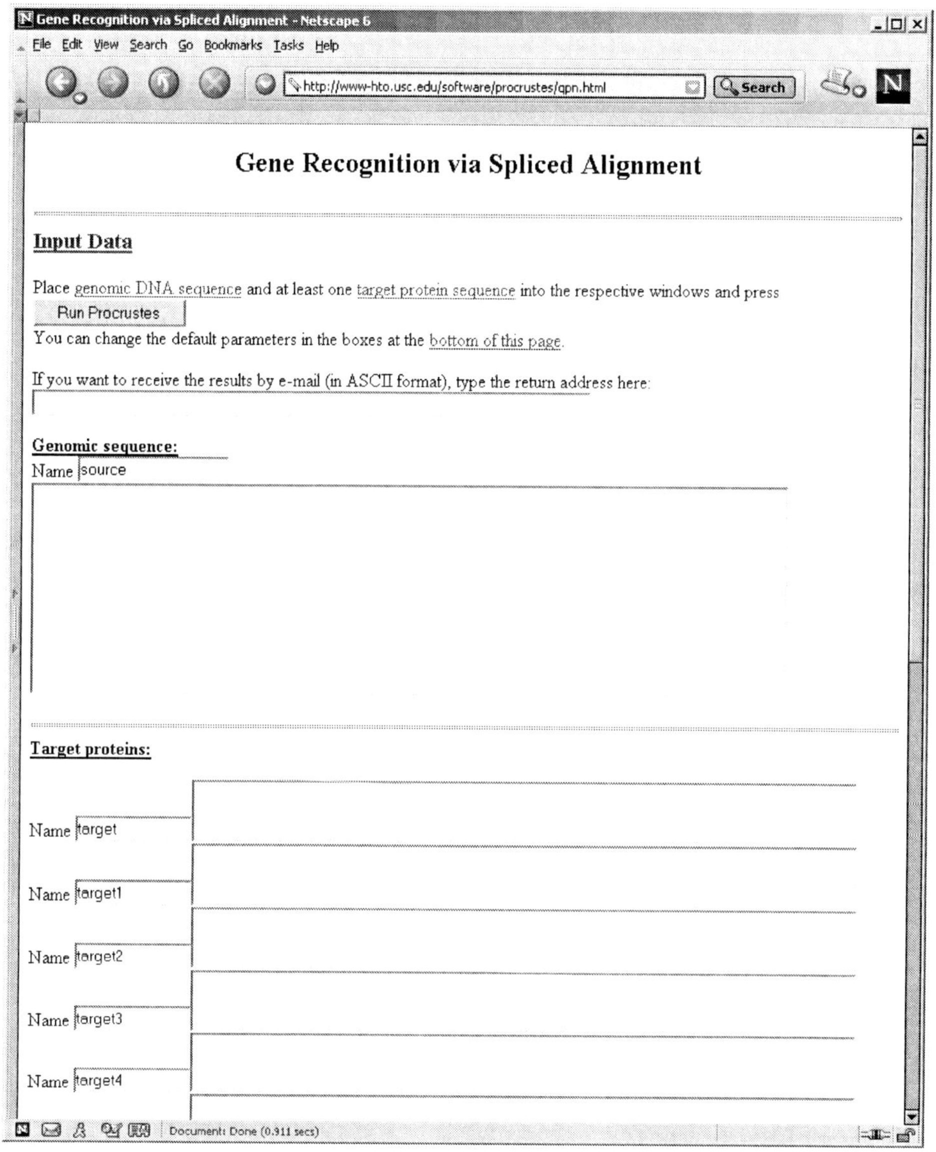

Figure 8.57a. The PROCRUSTES home page (part A). Available at http://www-hto.usc.edu/software/procrustes/wwwserv.html.

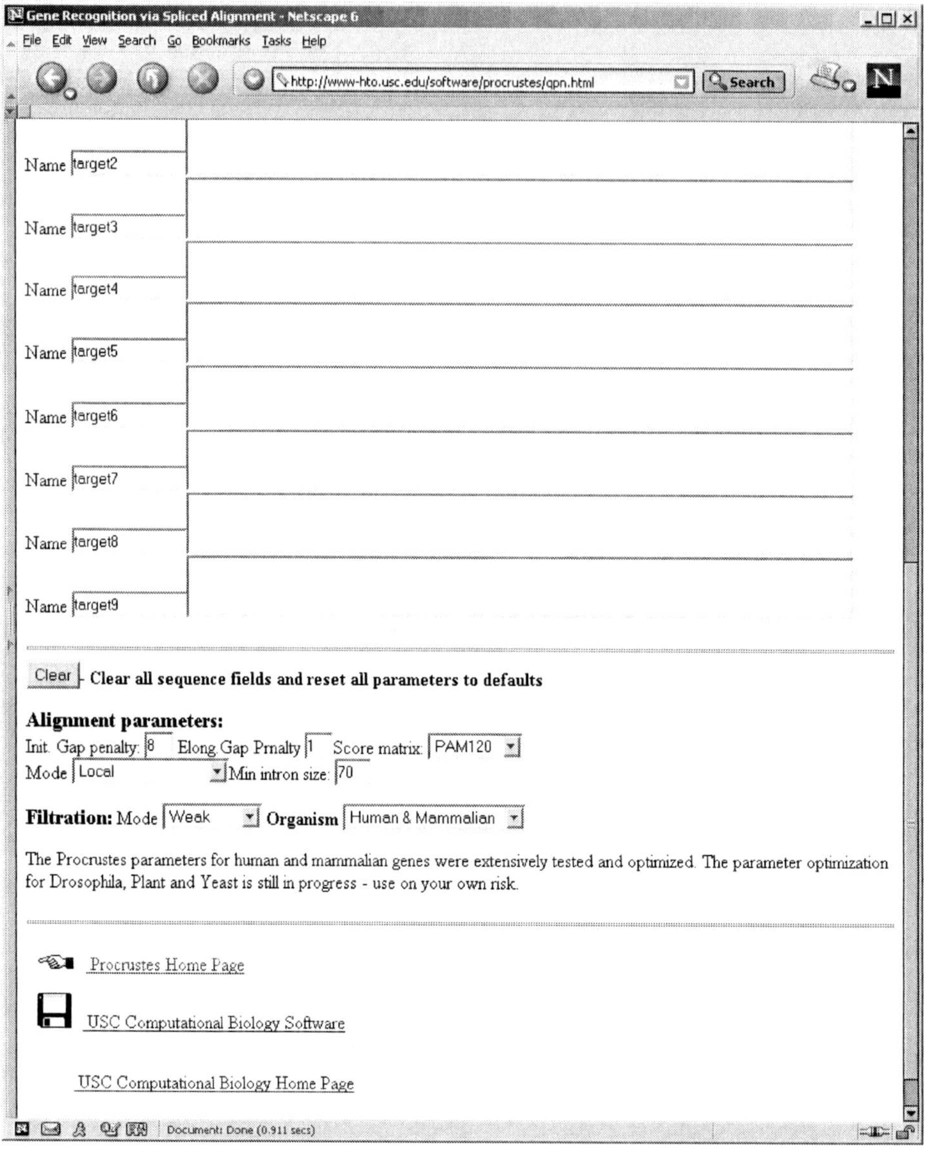

Figure 8.57b. The PROCRUSTES home page (part B). Available at http://www-hto.usc.edu/software/procrustes/wwwserv.html.

Output file

The output file consists, at the top, of an exon map where the location of the exon-intron regions is drawn for all the submitted sequences (the query and the related sequences). Thus, for each query-sequence/ related-sequence pair, three sections are prepared.

These sections are:

Target protein section. In this part of the output, information about the reliability of the prediction and the list of the putative exons are reported (Figure 8.58).

Translation of predicted gene. In this section the putative protein is displayed (Figure 8.59).

Spliced alignment. In this section the alignment between the query sequence and the related sequence is reported. The '*' symbol refers to an exact match, '|' identifies similar residues and '=' specifies exon boundaries (Figure 8.60).

```
Target Protein J04809
Spliced Alignment score=1011 (100.0%)

Correlation Coefficient 100.0%  Underprediction
0.0%
Prediction Quality      100.0%  Overprediction
0.0%

START  3982..3988   (    7)  —
  —    5534..5569   (   36)  —
  —    5742..5905   (  164)  —
  —    6656..6772   (  117)  —
  —   10075..10266  (  192)  —
  —   10508..10573  (   66)  STOP
```

Figure 8.58. The PROCRUSTES output file. Target protein section. In this part of the output, information about the reliability of the prediction and the list of the putative exons are reported.

```
Translation of Predicted Gene
1    MEEKLKKTKI IFVVGGPGSG KGTQCEKIVQ KYGYTHLSTG DLLRSEVSSG SARGKKLSEI
61   MEKGQLVPLE TVLDMLRDAM VAKVNTSKGF LIDGYPREVQ QGEEFERRIG QPTLLLYVDA
121  GPETMTQRLL KRGETSGRVD DNEETIKKRL ETYYKATEPV IAFYEKRGIV RKVNAEGSVD
181  SVFSQVCTHL DALK
```

Figure 8.59. The PROCRUSTES output file. Translation of predicted gene. In this section the putative protein is displayed.

```
Spliced Alignment with J04809

target    1 ME=LKKTKIIFVVGG=PGSGKGTQCEKIVQKYGYTHLSTGDLLRSEVSSGSARGKKLSEI
            ** =*************** =*****************************************
source    1 ME=LKKTKIIFVVGG=PGSGKGTQCEKIVQKYGYTHLSTGDLLRSEVSSGSARGKKLSEI

target   59 MEKGQLVPLET=VLDMLRDAMVAKVNTSKGFLIDGYPREVQQGEEFERRIG=QPTLLLYV
            *********** =*************************************** =********
source   59 MEKGQLVPLET=VLDMLRDAMVAKVNTSKGFLIDGYPREVQQGEEFERRIG=QPTLLLYV

target  117 DAGPETMTQRLLKRGETSGRVDDNEETIKKRLETYYKATEPVIAFYEKRGIVRKVN=AEG
            ********************************************************** =***
source  117 DAGPETMTQRLLKRGETSGRVDDNEETIKKRLETYYKATEPVIAFYEKRGIVRKVN=AEG

target  176 SVDSVFSQVCTHLDALK
            *****************
source  176 SVDSVFSQVCTHLDALK
```

Figure 8.60. The PROCRUSTES output file. Spliced alignment. In this section the alignment between the query sequence and the related sequence is reported. The '*' symbol refers to an exact match, '|' identifies similar residues and '=' specifies exon boundaries.

8.4 References

Altschul, S.F., Gish, W., Miller, W., Myers, E.W., and Lipman, D.J. 1990. Basic local alignment search tool. J. Mol. Biol. 215: 403-410.

Altschul, S.F. 1991. Amino acid substitution matrices from an information theoretic perspective. J. Mol. Biol. 219: 555-565.

Altschul, S.F. and Gish, W. 1996. Local alignment statistics. Meth. Enzymol. 266: 460-480.

Besemer, J. and Borodovsky, M. 1999. Heuristic approach to deriving models for gene finding. Nucl. Acids Res. 27: 3911-3920.

Besemer, J., Lomsadze, A., and Borodovsky, M. 2001. GeneMarkS: a self-training method for prediction of gene starts in microbial genomes. Implication for finding sequence motifs in regulatory regions. Nucl. Acids Res. 29: 2607-2618.

Borodovsky, M. and McIninch, J. 1993. GeneMark: parallel gene recognition for both DNA strands. Computers and Chemistry. 17: 123-133.

Burge, C., and Karlin, S. Prediction of complete gene structures in human genomic DNA. 1997. J. Mol. Biol. 268: 78-94.

Fortna, A., and Gardiner, K. 2001. Genomic sequence analysis tools: a user's guide. Trends Genet. 17: 158-164.

Gelfand, M.S., Mironov, A.A., and Pevzner, P.A. 1996. Gene recognition via spliced sequence alignment. Proc. Natl. Sci. USA. 93: 9061-9066.

Ghosh, D. 2000. Object-oriented transcription factors database (ooTFD). Nucl. Acids Res. 28: 308-310.

Guigo, R., Knudsen, S., Drake, N., and Smith, T. 1992. Prediction of gene structure. J. Mol. Biol. 226: 141-157.

Jurka, J., Klonowski, P., Dagman, V., and Pelton, P. 1996. CENSOR, a program for identification and elimination of repetitive elements from DNA sequences. Comput Chem. 20: 119-21.

Karlin, S. and Altschul, S.F. 1990. Methods for assessing the statistical significance of molecular sequence features by using general scoring schemes. Proc. Natl. Acad. Sci. USA. 87: 2264 - 2268.

Lipman, D.J., and Pearson, W.R., 1985. Rapide and sensitive protein similarity searches. Science. 227: 1435-1441.

Lukashin, A. and Borodovsky, M. 1998. GeneMark.hmm: new solutions for gene finding. Nucl. Acids Res. 26: 1107-1115.

Milanesi, L., Kolchanov, N.A., Rogozin, I.B., Ischenko, I.V., Kel, A.E., Orlov, Yu L., Ponomarenko, MP, and Vezzoni, P. 1993a. GenView: a computing tool for protein-coding regions prediction in nucleotide sequences.In: Proceedings of the Second International Conference on Bioinformatics, Supercomputing and Complex Genome Analysis (H.A. Lim, J.W. Fickett, C.R. Cantor and R.J. Robbins, eds.), World Scientific Publishing, Singapore, pp. 573-588.

Milanesi, L., Kolchanov, N., Rogozin, I., Kel, A, and Titov, I. 1993b. Sequence functional inference. In: Guide to human genome computing, ed. M.J.Bishop, Academic Press limited, Cambridge.

249-312.

Milanesi, L., Arrigo, P., and Muselli, M. 1995. Recognition of poly-A signals with hamming clustering. In: Proceedings of the Third International Conference on Bioinformatics, Supercomputing and Complex Genome Analysis (H.A. Lim, J.W. Fickett, C.R. Cantor and R.J. Robbins, eds.), World Scientific Publishing, Singapore, pp. 461-466.

Milanesi, L., Muselli, M., and Arrigo, P. 1996. Hamming Clustering method for signals prediction in 5' and 3' regions of eukaryotic genes. Comput. Applic. Biosci. 12: 399-404.

Milanesi, L., and Rogozin, I.B. 1998. Prediction of human gene structure. In: Guide to Human Genome Computing (2nd ed.) (Ed. M.J.Bishop) Academic Press, Cambridge. 215-259.

Milanesi, L., D'Angelo, D., and Rogozin, I.B. 1999. GeneBuilder: interactive in silico prediction of genes structure. Bioinformatics. 15: 612-621.

Needleman, S.B., and Wunsch., C.D. 1970. A general method applicable to the search for similarities in the amino acid sequence of two proteins. J. Mol. Biol. 48: 443-453.

Pearson, W.R., and Lipman, D.J., 1988. Improved tools for biological sequence comparison. Proc. Natl. Acad. Sci. U.S.A. 85: 2444-2448.

Perier, R.C., Praz, V., Junier, T., Bonnard, C., and Bucher, P. 2000. The eukaryotic promoter database (EPD). Nucl. Acids Res. 28: 302-303.

Prestridge, D.S. 1995. Predicting polII promoter sequences using transcription factor binding sites. J. Mol. Biol. 249: 923-932.

Rogozin, I.B., Kolchanov, N.A., and Milanesi L. 1995. A computing system for protein-coding regions prediction in Diptera nucleotide sequences. Drosophila Information Service. 76: 185-187.

Rogozin, I.B., Milanesi, L., and Kolchanov, N.A. 1996. Gene structure prediction using information on homologous protein sequence. Comput. Applic. Biosci. 12: 161-170.

Rogozin, I.B., and Milanesi, L. 1997. Analysis of donor splice signals in different organisms. J. Mol. Evol. 45: 50-59.

Rogozin, I.B., D'Angelo, D., and Milanesi, L. 1999. Protein coding regions prediction combining similarity searches and conservative evolutionary properties of protein coding sequences. Gene. 226: 129-137.

Rogozin, I.B., Kochetov, A.V., Kondrashov, F.A., Koonin, E.V., and Milanesi, L. 2001. Presence of ATG triplets in 5' untranslated regions of eukaryotic cDNAs correlates with a "weak" context of the start codon. Bioinformatics. 17: 890-900.

Shmatkov, A.M., Melikyan, A.A., Chernousko, F.L., and Borodovsky, M. 1999. Finding prokaryotic genes by the 'frame-by-frame' algorithm: targeting gene starts and overlapping genes. Bioinformatics. 15: 874-86.

Smith, T.F., and Waterman, M.S. 1981. Identification of common molecular subsequences. J. Mol. Biol. 147: 195-197.

Smith, T.F., Waterman, M.S., and Burks, C. 1985. The statistical distribution of nucleic acid similarities. Nucl. Acids Res. 13: 645-656.

Solovyev, V.V., Salamov, A.A., and Lawrence, C.B. 1994. Predicting internal exons by oligonucleotide composition and discriminant analysis of spliceable open reading frames. Nucl. Acids Res. 22: 5156-5163.

Solovyev, V.V., and Salamov, A.A. 1997. The Gene-Finder computer tools for analysis of human and model organisms genome sequences. In: Proceedings of the Fifth International Conference on Intelligent Systems for Molecular Biology (eds. Rawling C., Clark D., Altman R., Hunter, L., Lengauer, T., Wodak, S.), Halkidiki, Greece, AAAI Press, 294-302.

9

Practical Aspects of Protein Sequence Analysis

Allegra Via

Contents

From: *The Internet for Cell and Molecular Biologists: Current Applications and Future Potential*
ISBN 1-898486-32-8 © 2002 Horizon Scientific Press, Wymondham, UK

9.5 Motifs and patterns
    9.5.1 Pattern and domain databases
        9.5.1.1 PROSITE
        9.5.1.2 BLOCKS
        9.5.1.3 PFam
        9.5.1.4 PRINTS
    9.5.2 Servers for patterns and domains databases scanning
        9.5.2.1 ProfileScan
        9.5.2.2 BLOCKS server
        9.5.2.3 SMART server
9.6 References

Abstract

This chapter is dedicated to the analysis of amino acid sequences. It is organized in five subsections. In the first and second the reader can find a user-friendly description of sequence databases and instructions to use some of the main tools for pair-wise alignments and database searches. Section 9.3 is dedicated to multiple alignments while section 9.4 is a very short introduction to Hidden Markov Models. Finally, section 9.5 is an overview of the most important pattern and domain databases and describes tools to use them for protein sequence analysis.

Given one or a set of sequences you can essentially perform:

1.   Database searches looking for identical or similar sequences (for the detection of homology in the context of phylogenetic analysis and/or inference of function).
    For these purposes sections 9.1 and 9.2 provides a description of the most widely used protein sequence databases and tools (programs and servers) for searches in such databases.

    For this analysis we suggest the following steps:

    •   identify the most suitable database for your needs;
    •   select the most appropriate searching program

- **perform your search.**

The results of your search may be more or less biologically relevant. You can influence relevance and reliability by modifying the parameters of the searching program. If you do not feel self-confident in handling program parameters, we suggest using the default ones provided by the program itself.

2. **A multiple alignment.**
   (a) **one can align a single sequence to a multiple alignment of sequences provided by databases of protein families.**
   (b) **one can build a multiple alignment starting from a new set of sequences.**
   You can find the tools for both these in section 9.3.

3. **Pattern matching.**
   You may be interested in the identification of functional sites in a protein sequence (phosphorylation sites, glycosylation sites, etc.).
   Section 9.5 provides a description of databases and tools for the identification of biologically relevant signatures in protein sequences.

Many of the programs described in this section can be used directly through the WWW. Others can be downloaded from the suitable web site and installed on a local computer.

9.1 Protein sequence databases

The most reliable protein sequences databases are SWISSPROT-TrEMBL and PIR (Table 9.1). They are both non-redundant, expertly annotated and cross-referenced. However the SWISSPROT database contains a smaller number of entries (the SWISSPROT current release 40.0 of October 18, 2001, contains 101602 entries) but better annotated than the PIR entries (the current PIR release 70.01 of October 22, 2001, contains 254293 entries).

Table 9.1. WWW addresses of SWISSPROT-TREMBL and PIR sequence databases.

SWISSPROT-tremble (ExPASy)	http://www.expasy.ch/sprot/
SWISSPROT-TrEMBL (EBI)	http://www.ebi.ac.uk/swissprot/
PIR	http://pir.georgetown.edu/pirwww/search/textpsd.shtml

9.1.1 Swissprot-TrEMBL

SWISSPROT (Bairoch *et al.*, 2000) is a database of fully annotated protein sequences while TrEMBL contains all the translations of EMBL nucleotide sequence entries (see chapter 8) that are not yet annotated by the SWISSPROT staff. It is established and maintained by the Swiss Institute for Bioinformatics (SIB) and the European Bioinformatics Institute (EBI) whose molecular biology server WWW addresses are given in Table 9.2.

The format of SWISSPROT is as similar as possible to that of the EMBL Nucleotide Sequence Database (see chapter 8).

The SWISSPROT-TrEMBL home page at EBI is reported in Figure 9.1.

To retrieve a protein entry you can type the protein SWISSPROT ID or the AC (ACcession number) in the 'Quick search' window. If the sequence ID is unknown, a suitable keyword indicating the protein description, gene name or organism can be typed instead. This produces a list of SWISSPROT protein codes related to that keyword. A SWISSPROT entry is illustrated in Figure 9.2. Each entry corresponds to a single protein and contains the sequence data, the bibliographical references and the taxonomic data. The annotation of each entry gives information on the function of the protein, on domains and sites, on secondary, tertiary and quaternary structure, on similarity to others

Table 9.2. www addresses of SIB (Swiss Institute for Bioinformatics) and EBI (European Bioinformatics Institute) home pages.

SIB home page	http://www.isb-sib.ch/
EBI home page	http://www.ebi.ac.uk/

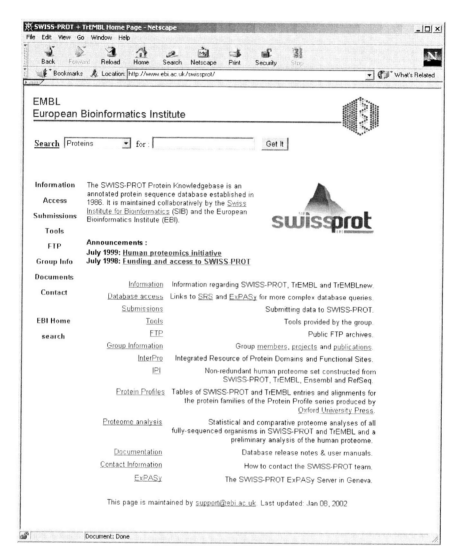

Figure 9.1. The SWISSPROT-TrEMBL home page at EBI.

proteins, on sequence conflict(s), on mutation(s), on post-translational modification(s), on disease(s) associated with deficiencie(s) in the protein. All this information is recorded in the different types of annotation lines that, together with the sequence lines, make up the SWISSPROT entry.

In Table 9.3 there is a short description of the two-character codes used at the start of entry lines. Each code indicates the type of data contained in the corresponding line.

Table 9.3. Short description of the two-character codes used to begin the entry lines of a SWISSPROT entry.

Line code	Content	Occurrence in an entry
ID	IDentification	Once; starts the entry
AC	ACcession number(s)	One or more
DT	DaTe	Three times
DE	DEscription	One or more
GN	Gene Name(s)	Optional
OS	Organism Species	One or more
OG	OrGanelle	Optional
OC	Organism Classification	One or more
RN	Reference Number	One or more
RP	Reference Position	One or more
RC	Reference Comment(s)	Optional
RX	Reference cross-reference(s)	Optional
RA	Reference Authors	One or more
RT	Reference Title	Optional
RL	Reference Location	One or more
CC	Comments or notes	Optional
DR	Database cross-references	Optional
KW	KeyWords	Optional
FT	Feature Table 9.data	Optional
SQ	SeQuence header	Once
	(blanks) sequence data	One or more
//	Termination line	Once; ends the entry

```
ID   143B_HUMAN      STANDARD;      PRT;    245 AA.
AC   P31946;
DT   01-JUL-1993 (Rel. 26, Created)
DT   01-FEB-1996 (Rel. 33, Last sequence update)
DT   01-MAR-2002 (Rel. 41, Last annotation update)
DE   14-3-3 PROTEIN BETA/ALPHA (PROTEIN KINASE C INHIBITOR PROTEIN-1)
DE   (KCIP-1) (PROTEIN 1054).
GN   YWHAB.
OS   Homo sapiens (Human).
OC   Eukaryota; Metazoa; Chordata; Craniata; Vertebrata; Euteleostomi;
OC   Mammalia; Eutheria; Primates; Catarrhini; Hominidae; Homo.
OX   NCBI_TaxID=9606;
RN   [1]
RP   SEQUENCE FROM N.A.
RC   TISSUE=KERATINOCYTES;
RX   MEDLINE=93294871; PubMed=8515476;
RA   Leffers H., Madsen P., Rasmussen H.H., Honore B., Andersen A.H.,
RA   Walbum E., Vandekerckhove J., Celis J.E.;
RT   "Molecular cloning and expression of the transformation sensitive
RT   epithelial marker stratifin. A member of a protein family that has
RT   been involved in the protein kinase C signalling pathway.";
```

Figure 9.2. WWW page of the SWISSPROT-TrEMBL SWALL:143B_HUMAN entry at EBI

```
RL    J. Mol. Biol. 231:982-998(1993).
RN    [ 2]
RP    SEQUENCE FROM N.A.
RA    Deloukas P., et al.
RT    «The DNA sequence and comparative analysis of human chromosome
20.»;
RL    Nature 414:865-871(2001).
RN    [ 3]
RP    SEQUENCE FROM N.A.
RC    TISSUE=SKIN;
RA    Strausberg R.;
RL    Submitted (DEC-2000) to the EMBL/GenBank/DDBJ databases.
CC    -!- FUNCTION: ACTIVATES TYROSINE AND TRYPTOPHAN HYDROXYLASES IN
THE
CC       PRESENCE OF CA(2+)/CALMODULIN-DEPENDENT PROTEIN KINASE II, AND
CC    STRONGLY ACTIVATES PROTEIN KINASE C. IS PROBABLY A MULTIFUNCTIONAL
CC    REGULATOR OF THE CELL SIGNALING PROCESSES MEDIATED BY BOTH    CC    -
!- SIMILARITY: BELONGS TO THE 14-3-3 FAMILY.
DR    EMBL; X57346; CAA40621.1; -.
DR    EMBL; AL008725; CAA15497.1; -.
DR    EMBL; BC001359; AAH01359.1; -.
DR    HSSP; P29312; 1A38.
DR    MIM; 601289; -.
DR    InterPro; IPR000308; 14-3-3.
DR    Pfam; PF00244; 14-3-3; 1.
DR    PRINTS; PR00305; 1433ZETA.
DR    ProDom; PD000600; 14-3-3; 1.
DR    SMART; SM00101; 14_3_3; 1.
DR    PROSITE; PS00796; 1433_1; 1.
DR    PROSITE; PS00797; 1433_2; 1.
KW    Brain; Neurone; Phosphorylation; Acetylation; Multigene family;
KW    Alternative initiation.
FT    INIT_MET       0       0      BY SIMILARITY.
FT    CHAIN          1     245      14-3-3 PROTEIN BETA/ALPHA, LONG ISOFORM.
FT    CHAIN          2     245      14-3-3 PROTEIN BETA/ALPHA,SHORT ISOFORM.
FT    INIT_MET       2       2      FOR SHORT ISOFORM (BY SIMILARITY).
FT    MOD_RES        1       1      ACETYLATION (BY SIMILARITY).
FT    MOD_RES        2       2      ACETYLATION (IN SHORT ISOFORM)
FT                                  (BY SIMILARITY).
FT    MOD_RES      185     185      PHOSPHORYLATION (BY SIMILARITY).
SQ    SEQUENCE   245 AA;  27951 MW;   0BCA59BF97595485 CRC64;
      TMDKSELVQK AKLAEQAERY DDMAAAMKAV TEQGHELSNE ERNLLSVAYK NVVGARRSSW
      RVISSIEQKT ERNEKKQQMG KEYREKIEAE LQDICNDVLE LLDKYLIPNA TQPESKVFYL
      KMKGDYFRYL SEVASGDNKQ TTVSNSQQAY QEAFEISKKE MQPTHPIRLG LALNFSVFYY
      EILNSPEKAC SLAKTAFDEA IAELDTLNEE SYKDSTLIMQ LLRDNLTLWT SENQGDEGDA
      GEGEN
//
```

Figure 9.2. Continued.

The **ID line** has the following general form:

ID ENTRY_NAME DATA_CLASS; MOLECULE_TYPE; SEQUENCE_LENGTH.

The ENTRY_NAME consists of up to ten uppercase alphanumeric characters and it is used to identify a sequence. The DATA_CLASS term serves to distinguish between fully annotated entries and those in TrEMBL. There are two possible DATA_CLASS: STANDARD refers to completely annotated sequences while PRELIMINARY refers to sequences that have not yet been annotated (i.e.TrEMBL entries). The MOLECULE_TYPE item indicates the type of molecule of the entry (in SWISSPROT PRT stands for PRoTein) and finally SEQUENCE_LENGTH is the number of the sequence residues.

Table 9.4. List of the possible DATABASE_IDENTIFIERS of a SWISSPROT entry, i.e. the abbreviated name of the cross-referenced database.

DATABASE IDENTIFIER	Description
EMBL	Nucleotide sequence database of EMBL (EBI)
CARBBANK	Complex carbohydrate structure database (CCSD) from CarbBank
DICTYDB	Dictyostelium discoideum genome database
ECO2DBASE	Escherichia coli gene-protein database (2D gel spots) (ECO2DBASE)
ECOGENE	Escherichia coli K12 genome database (EcoGene)
FLYBASE	Drosophila genome database (FlyBase)
GCRDB	G-protein--coupled receptor database (GCRDb)
HIV	HIV sequence database
HSC-2DPAGE	Harefield hospital 2D gel protein databases (HSC-2DPAGE)
HSSP	Homology-derived secondary structure of proteins database (HSSP)
INTERPRO	Integrated resource of protein families, domains and functional sites (InterPro)
MAIZEDB	Maize genome database (MaizeDB)
MAIZE-2DPAGE	Maize genome 2D Electrophoresis database (Maize-2DPAGE)
MENDEL	Plant gene nomenclature database (Mendel)
MGD	Mouse genome database (MGD)
MIM	Mendelian Inheritance in Man Database (MIM)
PDB	3D macromolecular structure Protein Data Bank (PDB)
PFAM	Pfam protein domain database (see ...)
PIR	Protein sequence database of the Protein Information Resource (PIR)
PRINTS	Protein Fingerprint database (PRINTS)
PROSITE	PROSITE protein domains and families database (see ...)
REBASE	Restriction enzyme database (REBASE)
AARHUS/GHENT-2DPAGE	Human keratinocyte 2D gel protein database from Aarhus and Ghent universities
SGD	Saccharomyces Genome Database (SGD)
STYGENE	Salmonella typhimurium LT2 genome database (StyGene)
SUBTILIST	Bacillus subtilis 168 genome database (SubtiList)
SWISS-2DPAGE	Human 2D Gel Protein Database from the University of Geneva (SWISS-2DPAGE)
TIGR	The bacterial database(s) of 'The Institute of Genome Research' (TIGR)
TRANSFAC	Transcription factor database (TRANSFAC)
TUBERCULIST	Mycobacterium tuberculosis H37Rv genome database (TubercuList)
WORMPEP	Caenorhabditis elegans genome sequencing project protein database (WormPep)
YEPD	Yeast electrophoresis protein database (YEPD)
ZFIN	Zebrafish Information Network genome database (ZFIN)

The **AC line** reports the list of accession numbers of a protein. The accession number is a reliable protein identifier always conserved from release to release. A single entry can have more than one accession number: when two ore more entries have been merged into one, the accession numbers of all the entries merged are listed in the AC line(s). Notice however that only the first one ('primary accession number') should be quoted in publications.

 The **DT lines** indicate the date of creation (first line), the date of the last sequence update (second line) and the date of the last annotation update (third line).

The **DE line(s)** are used to provide a general but short description of the protein.

The **GN line** lists the gene(s) name(s) that code for the protein. If an individual locus is associated to more than one gene name, all the synonyms are reported separated by the word 'OR'. If multiple genes code for an identical protein sequence, the different gene names are reported separeted by the word 'AND'.

The **OS line** reports the name(s) of the organism(s) from which the protein originated.

The **OG line** indicates the organelle (mitochondria, chloroplast, a cyanelle, or a plasmid) that is the source of the gene coding for a protein.

The **OC lines** report the NCBI taxonomic classification of the source organism. For more information about this type of taxonomic classification see http://www.ncbi.nlm.nih.gov/Taxonomy/.

The RN, RP, RC, RX, RA, RT, RL reference lines refer to literature quoted within SWISSPROT. A more detailed description of each of these lines can be found in Figure 9.2 with the help of the Table 9.3.

The CC lines contain comments or notes to the entry.

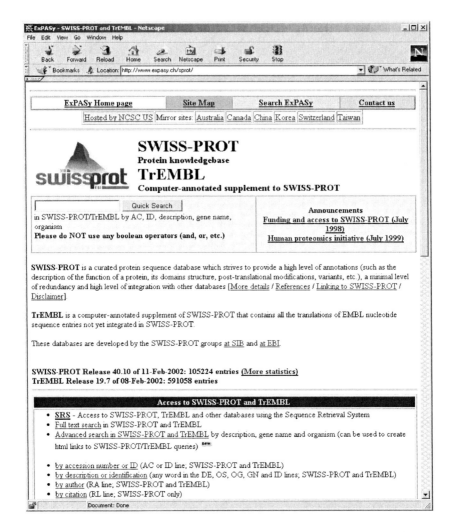

Figure 9.3. The www page of SWISSPROT-TrEMBL home page at ExPASy.

The DR lines list all the databases that contain information related to the SWISSPROT entry. The first item on the DR lines is the DATABASE_IDENTIFIER i.e. the abbreviated name of the cross-referenced database. The list of the possible DATABASE_IDENTIFIERS is reported in Table 9.4.

The **KW lines** contain word(s) that can be used for a literature search to retrieve information on the protein sequence.

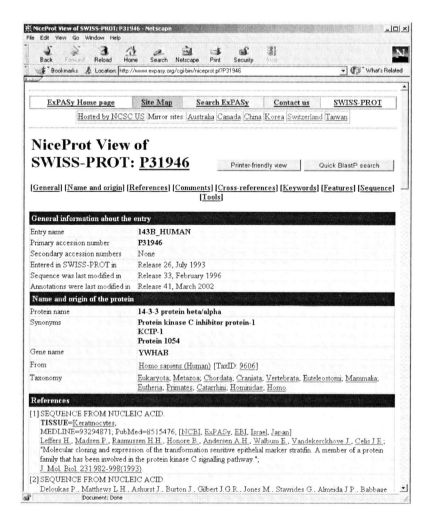

Figure 9.4. The www page of a NiceProt view of a SWISSPROT-TrEMBL entry (ExPASy).

The **FT lines** provide a feature table for the annotation of the sequence data. They list sequence regions or sites of interest such as binding sites, enzyme active sites etc.

The **SQ line** marks the beginning of the sequence data and provides a brief summary of its content.

The **sequence data lines** contain the string of amino acids of the protein

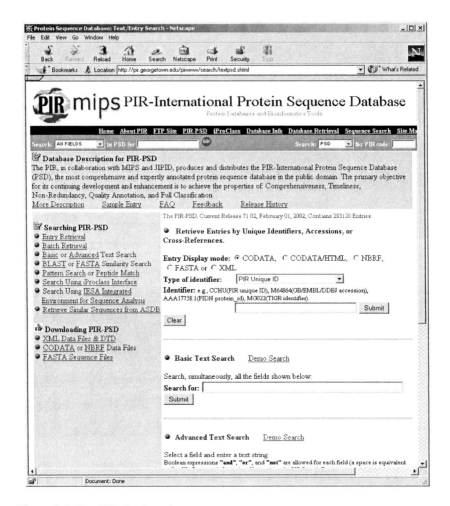

Figure 9.5. The PIR database home page.

sequence. Each line contains 60 amino acids separated in groups of 10. The amino acids are written in the standard IUPAC one letter code. The entry is ended by the '//' **termination line**.

The SWISSPROT-TrEMBL home page at ExPASy (SIB) can be accessed from the link provided in the SIB home page (Table 9.2) to the ExPASy proteomics server or by using the Web address reported in Table 9.1. The SWISSPROT-TrEMBL home page at ExPASy (SBI) is reported in Figure 9.3. As in the previous case you can use the 'Quick Search' window to perform your search (using ID or AC of a protein or a suitable keyword), obtaining the SWISSPROT code(s) which are linked to the NiceProt view of the SWISSPROT entry. The

NiceProt view of an entry contains the same information of the SWISSPROT entry but with the format shown in Figure 9.4.

9.1.2 PIR

The PIR protein sequence database (Barker *et al.*, 2001) is not as well annotated as SWISSPROT. However it is the most comprehensive protein sequence database in the public domain (~250,000 entries).

The PIR database can be reached from the PIR (Protein Information Resource) home page (http://pir.georgetown.edu/) or directly by typing or pasting the database WWW address (see Table 9.1) into a browser.

The PIR database homepage is shown in Figure 9.5. A protein can be retrieved using any identifier listed in the 'Type of identifier' menu and typing the protein ID in its text box. Otherwise a text search, both simple and advanced, can be performed. A typical PIR entry is shown in Figure 9.6. Some of the annotation and data lines are briefly explained in the caption of Figure 9.6. For a more complete description of a PIR entry see the 'Sample Entry' link in the PIR database homepage (whose WWW address is listed in Table 9.1).

9.2 Pair-wise alignments and database searches

In this section the most widely used tools, the FASTA, BLAST and PSI-BLAST programs, for pair-wise sequence alignments and database searches are described. In a similarity database search FASTA produces, as a result, the best alignment between the query sequence and the database sequences while BLAST provides, beyond the best alignment, many additional gapped alignments. Therefore, if you are interested in the similarity of a region of the query sequence with regions of the database sequences, BLAST is more suitable for your needs. However both these programs are well-tested and reliable indicators of sequence similarity. PSI-BLAST is more sensitive when searching distant relatives, with respect to FASTA and BLAST. However it can produce false positives by diverging from the original query sequence. Therefore its results should be carefully manually inspected.

```
ENTRY              S34755  #type complete
TITLE              14-3-3 protein (clone 1054) - human
ORGANISM           #formal_name Homo sapiens #common_name man
   #cross-references taxon:9606
DATE               31-Dec-1993 #sequence_revision 31-Dec-1993 #text_change
                   18-Jun-1999
ACCESSIONS         S34755; S29341
REFERENCE          S34753
  #authors         Leffers, H.; Madsen, P.; Rasmussen, H.H.; Honore, B.;
                   Andersen, A.H.; Walbum, E.; Vandekerckhove, J.; Celis,
                   J.E.
  #journal         J. Mol. Biol. (1993) 231:982-998
  #title           Molecular cloning and expression of the transformation
                   sensitive epithelial marker stratifin. A member of a
                   protein family that has been involved in the protein
                   kinase C signalling pathway.
  #cross-references MUID:93294871
  #accession       S34755
    ##status preliminary
    ##molecule_type mRNA
    ##residues 1-246 ##label LEF
    ##cross-references EMBL:X57346; NID:g23113; PIDN:CAA40621.1;
                   PID:g23114
CLASSIFICATION     #superfamily 14-3-3 protein
SUMMARY            #length 246 #molecular_weight 28082

SEQUENCE
                5         10        15        20        25        30
      1 M T M D K S E L V Q K A K L A E Q A E R Y D D M A A A M K A
     31 V T E Q G H E L S N E E R N L L S V A Y K N V V G A R R S S
     61 W R V I S S I E Q K T E R N E K K Q Q M G K E Y R E K I E A
     91 E L Q D I C N D V L E L L D K Y L I P N A T Q P E S K V F Y
    121 L K M K G D Y F R Y L S E V A S G D N K Q T T V S N S Q Q A
    151 Y Q E A F E I S K K E M Q P T H P I R L G L A L N F S V F Y
    181 Y E I L N S P E K A C S L A K T A F D E A I A E L D T L N E
    211 E S Y K D S T L I M Q L L R D N L T L W T S E N Q G D E G D
    241 A G E G E N
```

PDB structures most related to S34755:
   1A4OB (3-246) 87.3%

SCOP: 1A4O
CATH: 1A4O
FSSP: 1A4O
MMDB: 1A4O

Complex/Interaction Link for S34755:
   DIP: S34755

ALIGNMENTS containing S34755:
   FA2001 14-3-3 protein - 177.2 1.0
   M00871 14-3-3 protein - 836.0 1.0

Link to *i*ProClass (Superfamily classification and Alignment):
   *i*ProClass Report for S34755 at PIR.

Figure 9.6. The PIR entry of SWALL:143B_HUMAN.

The underlined items are linked to another WWW page.

The ENTRY line contains the PIR code of the protein sequence.

The TITLE line provides a short description of the protein.

The ORGANISM line(s) display three items:

#formal_name is followed by the formal name of the organism from which the sequence is obtained

#common_name is followed by the common name of the organism from which the sequence is obtained
Both the formal and the common names are linked to the list of PIR entries for the organism.

#cross-references is followed by the NCBI taxonomy ID which is linked to the taxomic classification organism page manteined at the NCBI. For more information about this type of taxonomic classification see http://www.ncbi.nlm.nih.gov/Taxonomy/.

The DATE line reports the date of the entry creation.

#sequence_revision indicates the date of the last revision of the sequence, while the #text_change indicates the date of the last annotation update.

ACCESSIONS line(s) report the list of the accession numbers for sequences merged into the entry.

REFERENCE line provides links to other PIR entries with same citation

#authors

#journal   citation for the protein

#title

#cross-references gives the link to medline abstract for this paper.

#accession protein reports the sequence accession number to which this paper refers

The following items refer to the sequence to which the paper refers:

##status indicates the status of the sequence

##molecule_type gives the type of molecule (DNA,RNA, protein)

##residues displays the first and the last residue number of the sequence

##label provides a link to the sequence

##cross-references provides links to other database entries for the sequence

CLASSIFICATION line

#superfamily provides the names of superfamily to which the protein sequence belongs. Each superfamily name is linked to the list of PIR entries classified into the same superfamily.

SUMMARY line

#length is followed by the number of the residues that make the sequence.

#molecular_weight of the protein

Figure 9.6. Continued.

Table 9.5. List of fastA programs available at EBI

PROGRAM	FUNCTION
fasta3	scans a protein or DNA sequence library for similar sequences
fastx/y3	compares a DNA sequence to a protein sequence database, comparing the translated DNA sequence in forward and reverse frames
Tfastx/y3	compares a protein to a translated DNA data bank
fasts3	compares linked peptides to a protein databank
fastf3	compares mixed peptides to a protein databank

Table 9.6. List of protein sequence databases available for the use of
Fasta3 program at EBI

swall	SWALL Non-Redundant Protein sequence database Swissprot+Trembl+TremblNew
swissprot	SWISS-PROT Protein Database
swnew	Updates to SWISS-PROT
sptremb	SPTREMBL (tremble)
remtrembl	REMTREMBL (uncurated entries in tremble)
prints	FingerPrints
SGT	Structural Genomic Targets Database
PDB	Protein Database of Brookhaven
IMGThla	IMGThla Immunogenetics Database

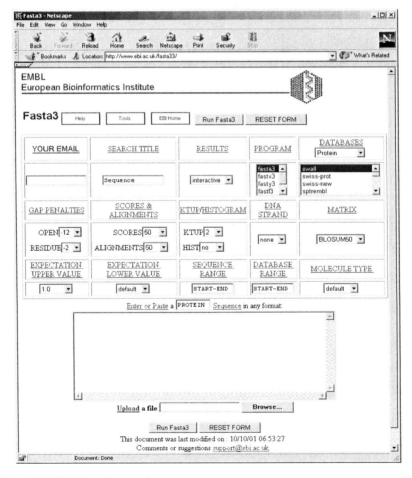

Figure 9.7. The Fasta3 www home page.

9.2.1 FASTA

FASTA at EBI (Pearson *et al.*, 1988; Pearson, 1990)

The list and function of fastA programs available at EBI is reported in Table 9.5. In the following section the use of Fasta3 is described. Fasta3 is a program for sequence database similarity searches whose home page is reported in Figure 9.7. Available databases for protein sequence searches are reported in Table 9.6. To submit a sequence the user must type or paste it into the text window in a suitable format. Fasta and free text (characters representing protein residues) formats are accepted. If you want to retrieve the fasta format of a protein sequence belonging to the SWISSPROT and/or PIR databases you can click on the FASTA format link at the bottom of the NiceProt view of the SWISS-PROT protein www entry page or on the FASTA link at the top of the PIR protein www entry page. An example of a sequence in FASTA format is given in Figure 9.8.

If you want to receive by email the result of your Fasta3 search, you must type or paste your address in the email text box and select the 'email' option, otherwise you will get the interactive run.

The Fasta3 program allows the choice of many parameters. The less 'intuitive' of them are: gap penalties, scores and alignments, *ktup*, matrix and expectation value (upper and lower).

In the following, these parameters will be discussed briefly, hopefully allowing the user to obtain a result that is as biologically correct as possible.

```
>sp|Q9ULV5|HSF4_HUMAN Heat shock factor protein 4 (HSF 4) (Heat
shock transcription factor 4) (HSTF 4) (hHSF4) - Homo sapiens
(Human).
MVQEAPAALPTEPGPSPVPAFLGKLWALVGDPGTDHLIRWSPSGTSFLVSDQSRFAKEVL
PQYFKHSNMASFVRQLNMYGFRKVVSIEQGGLLRPERDHVEFQHPSFVRGREQLLERVRR
KVPALRGDDGRWRPEDLGRLLGEVQALRGVQESTEARLRELRQQNEILWREVVTLRQSHG
QQHRVIGKLIQCLFGPLQAGPSNAGGKRKLSLMLDEGSSCPTPAKFNTCPLPGALLQDPY
FIQSPLPETNLGLSPHRARGPIISDIPEDSPSPEGTRLSPSSDGRREKGLALLKEEPASP
GGDGEAGLALAPNECDFCVTAPPPLPVAVVQAILEGKGSFSPEGPRNAQQPEPGDPREIP
DRGPLGLESGDRSPESLLPPMLLQPPQESVEPAGPLDVLGPSLQGREWTLMDLDMELSLM
QPLVPERGEPELAVKGLNSPSPGKDPTLGAPLLLDVQAALGGPALGLPGALTIYSTPESR
TASYLGPEASPSP
```

Figure 9.8. Example of FASTA format.

Results of Search:

Program: fasta33_t
Database: +swall+
Title: Sequence
SeqLen: 493

```
FASTA searches a protein or DNA sequence data bank
 version 3.3t09 May 18, 2001
Please cite:
 W.R. Pearson & D.J. Lipman PNAS (1988) 85:2444-2448

@:1-: 493 aa
 sp Heat shock factor protein 4 (HSF 4) (Heat shock transcription factor 4)
(HSTF 4) (hHSF4) - Homo sapiens (Human).
 vs   SWISS-PROT All library
searching /ebi/services/idata/fastadb/swall library

222471549 residues in 699662 sequences
 statistics extrapolated from 60000 to 698932 sequences
  Expectation_n fit: rho(ln(x))= 6.9303+/-0.0002; mu= 1.8719+/- 0.011
 mean_var=111.4450+/-21.889, 0's: 132 Z-trim: 63  B-trim: 457 in 2/63
 Lambda= 0.1215

FASTA (3.39 May 2001) function [optimized, BL50 matrix (15:-5)] ktup: 2
 join: 37, opt: 25, gap-pen: -12/ -2, width:  16
 Scan time: 10.990
```

```
The best scores are:                                       opt bits E(698932)
SWALL:HSF4_HUMAN Q9ULV5 HEAT SHOCK FACTOR PROTEIN  ( 493) 3384  604 2.6e-171
SWALL:HSF4_MOUSE Q9R0I1 HEAT SHOCK FACTOR PROTEIN  ( 492) 2889  517 3.4e-145
SWALL:Q9JIZ7 Q9JIZ7 HEAT SHOCK FACTOR 4.           ( 492) 2884  516 6.2e-145
SWALL:Q9JIZ8 Q9JIZ8 HEAT SHOCK FACTOR 4.           ( 492) 2872  514 2.7e-144
SWALL:HSF4_HUMAN Q9ULV5-01 SPLICE ISOFORM HSF4A O  ( 463) 1684  306 1.2e-81
SWALL:HSF4_MOUSE Q9R0I1-01 SPLICE ISOFORM HSF4A O  ( 462) 1575  287 6.8e-76
SWALL:HSF1_HUMAN Q00613-01 SPLICE SHORT ISOFORM O  ( 489)  932  174 6.1e-42
SWALL:AAH14638 AAH14638 SIMILAR TO HEAT SHOCK TRA  ( 529)  925  173 1.5e-41
SWALL:HSF1_HUMAN Q00613 HEAT SHOCK FACTOR PROTEIN  ( 529)  925  173 1.5e-41
SWALL:AAH13716 AAH13716 SIMILAR TO HEAT SHOCK FAC  ( 503)  921  172 2.4e-41
```

The first 10 best scoring sequences

```
>>SWALL:Q9JIZ8 Q9JIZ8 HEAT SHOCK FACTOR 4.                   (492 aa)
 initn: 2831 init1: 2831 opt: 2872  Z-score: 2728.1  bits: 514.3 E(): 2.7e-
144
Smith-Waterman score: 2872;  86.151% identity (86.327% ungapped) in 491 aa
overlap (2-492:1-490)

             10        20        30        40        50        60
sp    MVQEAPAALPTEPGPSPVPAFLGKLWALVGDPGTDHLIRWSPSGTSFLVSDQSRFAKEVL
      .::::::::::::::::::::::::::::::::::::::::::::::::::::::::::::::
SWALL:  MQEAPAALPTEPGPSPVPAFLGKLWALVGDPGTDHLIRWSPSGTSFLVSDQSRFAKEVL
        10        20        30        40        50

             70        80        90       100       110       120
sp    PQYFKHSNMASFVRQLNMYGFRKVVSIEQGGLLRPERDHVEFQHPSFVRGREQLLERVRR
      ::::::::::::::::::::::::::::::::::::::::::::::::::::::::::::::::
SWALL:  PQYFKHSNMASFVRQLNMYGFRKVVSIEQGGLLRPERDHVEFQHPSFVRGREQLLERVRR
          60        70        80        90       100       110
```

Figure 9.9. The Fasta3 output of the HSF4_HUMAN heat shock factor protein 4 of Figure 9.6. The underlined items are linked to the corresponding SWALL protein entries. Only the second alignment is reported. In the alignment lines, two vertical points mean identity, one point means similarity (the two aligned residues are both hydrophobic or charged etc.), no point (blank) means that the two aligned residues are different.

Gap penalties

For a general discussion of gap penalties see chapter 7.

The choice of gap parameters is empirical. However, since mutation events are rare but may involve several adjacent residues, it is common to use a *gap opening* value higher then a *gap extension* value. If the sequence you want to submit does not show any particular insertion and/or deletion requirements, the default gap penalties parameters (gap opening = -12 and gap extension = -2 for fasta with proteins) are usually recommended. On the other hand, if you wish to change these parameters, increase their value if you are looking for close homologues and decrease them if you need a more flexible search and you want to include distant homologues in the resulting matches.

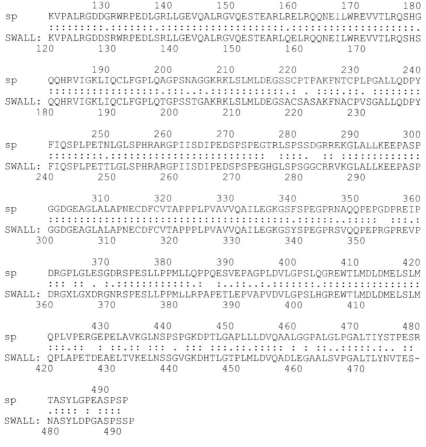

Figure 9.9. Continued.

This could however bias the results by increasing the number of non-homologous sequences with a high similarity score.

Scores and alignments

The user can select from a pull-down menu the maximum number of alignments reported in the output file.

Ktup

The *ktup* parameter controls the trade-off between speed and sensitivity. A high value parameter decreases the sensitivity of the search and

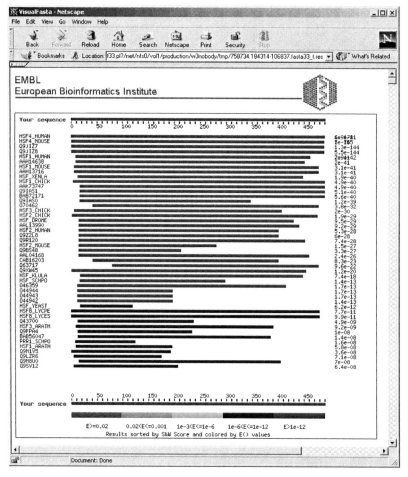

Figure 9.10. The VisualFasta image that precedes the ordinary Fasta3 output.

increases its speed. The default value for comparing proteins is 2, however, when searching for very distant relatives of a protein a lower value of *ktup* is preferred (for example *ktup* = 1).

Matrix

As explained above (see chapter 7) weight matrices cover evolutionary constraints. The most used substitution matrices are those based on the point-accepted-mutation (PAM) model of evolution (Dayhoff *et al.*, 1978) and those derived from the BLOCKS database (Henikoff and Henikoff., 1991) called BLOSUM matrices (Henikoff and Henikoff, 1992). If you are searching for distant relatives of your sequence, best results are obtained at higher PAM values (PAM250) and lower BLOSUM values (BLOSUM50). If, on the other hand, you are searching for sequences with a greater degree of similarity to your sequence, lower PAM values and higher BLOSUM values are to be preferred.

9.2.2 Fasta3 output

The Fasta3 output for the HSF4_HUMAN heat shock factor protein 4 of Figure 9.6 is partially reported in Figure 9.9. For editing reasons, the two values of SCORE and ALIGNMENT parameters have been chosen equal to 10. All the other parameter values are at their default.

The Fasta3 program provides also a 'VisualFasta' of the output (you can find the link at the top of the output page). It consists of the image reported in Figure 9.10, followed by the usual output page, as in Figure 9.9.

The FASTA program was the first program used for database similarity searches. However, it should be noted that it provides only the single optimum alignment for each database sequence. Therefore,

---

Table 9.7. Some FASTA www addresses.

FASTA at EBI	http://www.ebi.ac.uk/fasta33/
FASTA at the U. of Virginia	http://fasta.bioch.virginia.edu/
NCBI BLAST home page	http://www.ncbi.nlm.nih.gov/BLAST/

if the protein sequence contains multiple modules, some alignments can be missed.

As for many programs, it is possible to run Fasta from many different servers. Some of their WWW addresses are reported in Table 9.7.

9.2.3 BLAST

As of today, BLAST (Altschul *et al.*,1997) is the most widely used program for pair-wise sequence analysis and similarity searching. It is statistically more reliable than other programs and produces, beyond the best alignment, many additional gapped alignments, provided that they do not intersect those already reported.

The NCBI BLAST home page is shown in Figure 9.11. If you want to compare a protein sequence against a protein sequence database you must use the standard protein-protein BLAST program (blastp).

To perform your search you can type or paste your sequence in a suitable format in the text window of the blastp WWW page (see Figure 9.12).

You can use, as a suitable format, just lines of sequence data where sequences are expected to be represented in the standard IUB/IUPAC amino acid code (see Table 9.8).

Table 9.8. The standard IUB/IUPAC amino acid code

A	alanine	P	proline
B	aspartate or asparagine	Q	glutamine
C	cystine	R	arginine
D	aspartate	S	serine
E	glutamate	T	threonine
F	phenylalanine	U	selenocysteine
G	glycine	V	valine
H	histidine	W	tryptophan
I	isoleucine	Y	tyrosine
K	lysine	Z	glutamate or glutamine
L	leucine	X	any
M	methionine	*	translation st
N	asparagine	-	gap of indeterminate length

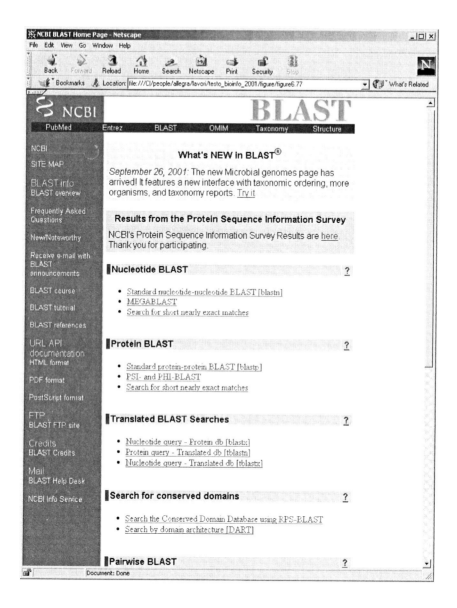

Figure 9.11. The NCBI BLAST home page.

The fasta format (Figure 9.8) and the sequence portion of a GenBank/GenPept flatfile report (Figure 9.13) also provide suitable formats for the BLAST program. The NCBI accession number of the protein or the gi (Genbank Identification) number of the gene codifying for the protein are also accepted as input.

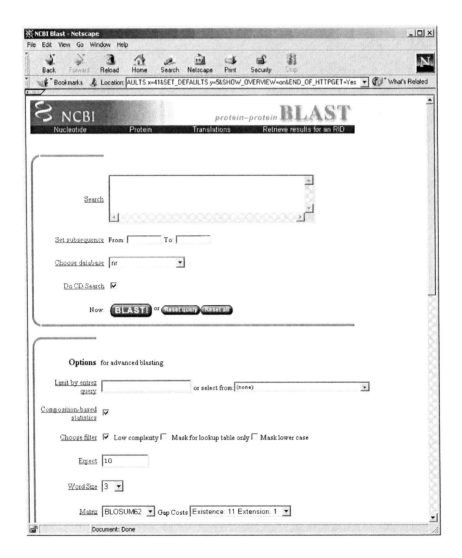

Figure 9.12a. The protein-protein BLAST (blastp) WWW page (part A).

In the 'Set subsequence' boxes you may enter a region of your query sequence and in the 'Choose database' box you can choose between the nr (all non-redundant GenBank CDS translations+PDB+ SwissProt+PIR+PRF) and the SWISSPROT databases.

At this point the BLAST program (blastp), the query sequence and the database are settled and the program can run (BLAST!). If you do not set any optional parameters (listed in the two sections:

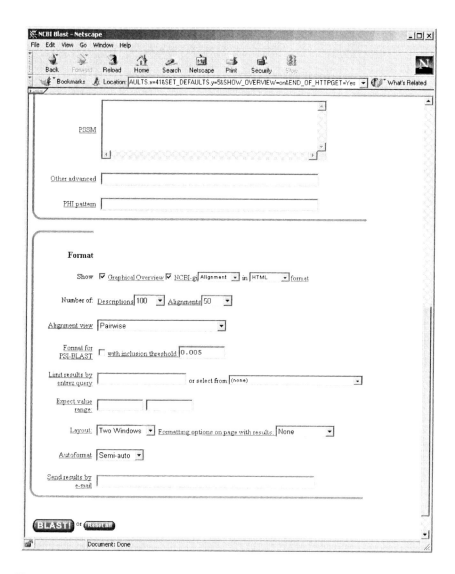

Figure 9.12b. The protein-protein BLAST (blastp) WWW page (part B).

'Options for advanced blasting' and 'Format') the program will run with the default parameters. You can find the meaning of each parameter plus various suggestions on its choice/use by clicking the corresponding link.

The most important of these parameters are discussed in the following section.

'Options for advanced blasting'

Filter (Low-complexity)

The default setting is 'ON' and it means that repetitive or low-complexity sequences (that might obscure more informative hits) are filtered by the program. You should turn off this filter if you are interested in a low complexity region of the query sequence. Moreover if the query is a short sequence and the number of hits of the search result is small, you can turn off this filter and search again.

Filter (Mask lower case)

The default setting is 'OFF". However if you want to mask a special region of your query sequence you can denote it with lower case characters, paste the rest of the sequence in upper case characters and turn on this filter.

Expect (E value)

The default value for this parameter is 10. It means that searching the database with a random query with similar length to your query, 10 hits with scores equal to or better than the defined alignment score are expected to occur by chance. Decreasing the E value produces a more stringent search, while a higher EXPECT threshold leads to a less stringent search. In particular, if the query is a short sequence, it can

```
  1 mvqeapaalp tepgpspvpa flgklwalvg dpgtdhlirw spsgtsflvs dqsrfakevl
 61 pqyfkhsnma sfvrqlnmyg frkvvsieqg gllrperdhv efqhpsfvrg reqllervrr
121 kvpalrgddg rwrpedlgrl lgevqalrgv qestearlre lrqqneilwr evvtlrqshg
181 qqhrvigkli qclfgplqag psnaggkrkl slmldegssc ptpakfntcp lpgallqdpy
241 fiqsplpetn lglsphrarg piisdipeds pspegtrlsp ssdgrrekgl allkeepasp
301 ggdgeaglal apnecdfcvt appplpvavv qailegkgsf spegprnaqq pepgdpreip
361 drgplglesg drspesllpp mllqppqesv epagpldvlg pslqgrewtl mdldmelslm
421 qplvpergep elavkglnsp spgkdptlga pllldvqaal ggpalglpga ltiystpesr
```

Figure 9.13. The sequence portion of a GenBank/GenPept flatfile report of the same sequence of Figure 9.8.

be found by chance many times in the database. Therefore, to take into account this possibility, one may increase the E value up to 1000 or more.

Word size

The word size default value is 3. You can reduce it to 2 in the case of a short query sequence to speed up the search.

Matrix

See the 'Matrix' subsection of the 'FASTA' section for a discussion of this parameter. In the case of BLAST, the default setting is BLOSUM62, which has been empirically shown to be among the best for detecting weak similarities.

Gap costs

See chapter 7 for a general discussion of these parameters. The BLAST default gap existence and gap extension penalty values are –11 and –1 respectively. If you are interested in hits that align to the entire length of the query sequence you can increase the gap existence penalty.

'Format'

Graphical overview, NCBI-gi and format

BLAST shows by default a graphical overview of the results. It is shown in Figure 9.14 and its use is discussed in the BLAST OUTPUT section (see below). In addition to the accession and/or locus name, BLAST allows the NCBI gi identifiers to be shown in the output.

HTML is the default format. However an alternative format may be selected.

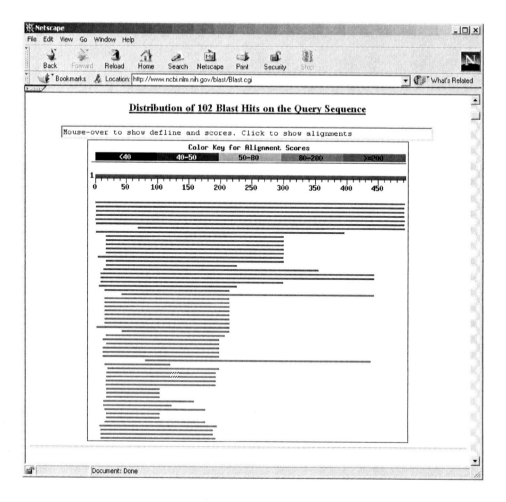

Figure 9.14. Graphical overvew of BLAST results.

Number of descriptions and alignments

The default settings are 100 and 50 respectively. This means that 100 descriptions and 50 alignments are shown by default. If the query sequence belongs to a large sequence family and you want to detect low-similarity hits, the number of descriptions and alignments should be increased. If you are looking only for sequences with high similarity to the query, you may reduce these numbers, speeding up the search.

```
BLASTP 2.2.1 [Apr-13-2001]

Reference:
Altschul, Stephen F., Thomas L. Madden, Alejandro A. Schäffer,
Jinghui Zhang, Zheng Zhang, Webb Miller, and David J. Lipman (1997),
„Gapped BLAST and PSI-BLAST: a new generation of protein database search
programs", Nucleic Acids Res. 25:3389-3402.

RID: 1002644571-20127-26268

Query= (480 letters)

Database: nr
          771,594 sequences; 245,249,562 total letters

If you have any problems or questions with the results of this search
please refer to the BLAST FAQs

Taxonomy reports
                                               Score      E
Sequences producing significant alignments:             (bits)  Value

gi|12643899|sp|Q9ULV5|HSF4_HUMAN  HEAT SHOCK FACTOR PROTEIN ...   759   0.0
gi|13124312|sp|Q9R0L1|HSF4_MOUSE  HEAT SHOCK FACTOR PROTEIN ...   679   0.0
gi|8886177|gb|AAF80399.1|AF160966_1  (AF160966) heat shock f...   677   0.0
gi|8886175|gb|AAF80398.1|  (AF160965) heat shock factor 4 [M...   674   0.0
gi|4557651|ref|NP_001529.1|  heat shock transcription factor...   613   e-174
gi|6754252|ref|NP_036069.1|  heat shock transcription factor...   548   e-155
gi|5921133|dbj|BAA84581.1|  (AB029347) transcription factor ...   499   e-140
gi|13647383|ref|XP_007871.2|  heat shock transcription facto...   487   e-137
gi|14861594|gb|AAK73747.1|AF391099_1  (AF391099) heat shock ...   248   9e-65
gi|8117742|gb|AAF72750.1|AF159134_1  (AF159134) heat shock t...   248   1e-64
```

description lines

```
>gi|13124312|sp|Q9R0L1|HSF4_MOUSE HEAT SHOCK FACTOR PROTEIN 4 (HSF 4) (HEAT SHOCK TRANSCRIPTION
           FACTOR 4) (HSTF 4) (MHSF4)
  gi|5921137|dbj|BAA84583.1| (AB029349) transcription factor HSF4b isoform [Mus musculus]
           Length = 492

 Score = 679 bits (1751), Expect = 0.0
 Identities = 380/478 (79%), Positives = 395/478 (82%)

Query: 2    VQEAPAALPTEPGPSPVPAFLGKLWALVGDPGTDHLIRWSPSGTSFLVSDQSRFAKEVLP 61
            +QEAPAALPTEPGPSPVPAFLGKLWALVGDPGTDHLIRWSPSGTSFLVSDQSRFAKEVLP
Sbjct: 1    MQEAPAALPTEPGPSPVPAFLGKLWALVGDPGTDHLIRWSPSGTSFLVSDQSRFAKEVLP 60

Query: 62   QYFKHSNMASFVRQLNMYGFRKVVSIEQGGLLRPERDHVEFQHPSFVRGREQLLERVRRK 121
            QYFKHSNMASFVRQLNMYGFRKVVSIEQGGLLRPERDHVEFQHPSFVRGREQLLERVRRK
Sbjct: 61   QYFKHSNMASFVRQLNMYGFRKVVSIEQGGLLRPERDHVEFQHPSFVRGREQLLERVRRK 120

Query: 122  VPALRGDDGRWRPEDLGRLLGEVQALRGVQESTEARLRELRQQNEILWREVVTLRQSHGQ 181
            VPALRGDD RWRPEDL RLLGEVQALRGVQESTEARL+ELRQQNEILWREVVTLRQSH Q
Sbjct: 121  VPALRGDDSRWRPEDLSRLLGEVQALRGVQESTEARLQELRQQNEILWREVVTLRQSHSQ 180

Query: 182  QHRVIGKLIQCLFGPLQAGPSNAGGKRKLSLMLDEGSSCPTPAKFNTCPLPGALLQDPYF 241
            QHRVIGKLIQCLFGPLQ GPS+ G KRKLSLMLDEGS+C  AKFN CP+ GALLQDPYF
Sbjct: 181  QHRVIGKLIQCLFGPLQTGPSSTGAKRKLSLMLDEGSACSASAKFNACPVSGALLQDPYF 240

Query: 242  IQSPLPETNLGLSPHRARGPIISDIPEDSPSPEGTRLSPSSDGRREKGLALLKEEPASPG 301
            IQSPLPET LGLSPHRARGPIISDIPEDSPSPEG RLSPS   RR KGLALLKEEPASPG
Sbjct: 241  IQSPLPETTLGLSPHRARGPIISDIPEDSPSPEGHRLSPSGGCRRVKGLALLKEEPASPG 300

Query: 302  GDGEAGLALAPNECDFCXXXXXXXXXXXXXXXXILEGKGSFSPEGPRNAQQPEPGDPREIPD 361
            GDGEAGLALAPNECDFC            ILEGKGS+SPEGPR+ QQPEP PRE+PD
Sbjct: 301  GDGEAGLALAPNECDFCVTAPPPLPVAVVQAILEGKGSYSPEGPRSVQQPEPRGPREVPD 360

Query: 362  RGPLGLESGDRXXXXXXXXXXXXXXXXXXXXVEPAGPLDVLGPSLQGREWTLMDLDMELSLMQ 421
            RG LGL+ G+R          +EP  P+DVLGPSL GREWTLMDLDMELSLMQ
Sbjct: 361  RGTLGLDRGNRSPESLLPPMLLRPAPETLEPVAPVDVLGPSLHGREWTLMDLDMELSLMQ 420

Query: 422  PLVPERGEPELAVKGLNSPSPGKDPTLGAPLLLDVQXXXXXXXXXXXXXXXTIYSTPES 479
            PL PE  E EL VK LNS  GKD TLG PL+LDVQ              T+Y+  ES
Sbjct: 421  PLAPETDEAELTVKELNSSGVGKDHTLGTPLMLDVQADLEGAALSVPGALTLYNVTES 478
```

Figure 9.15. The partial results of BLAST for the sequence of Figure 9.13. Ten descriptions and only the second highest scoring alignment are reported. The underlined items are linked to the corresponding www page.

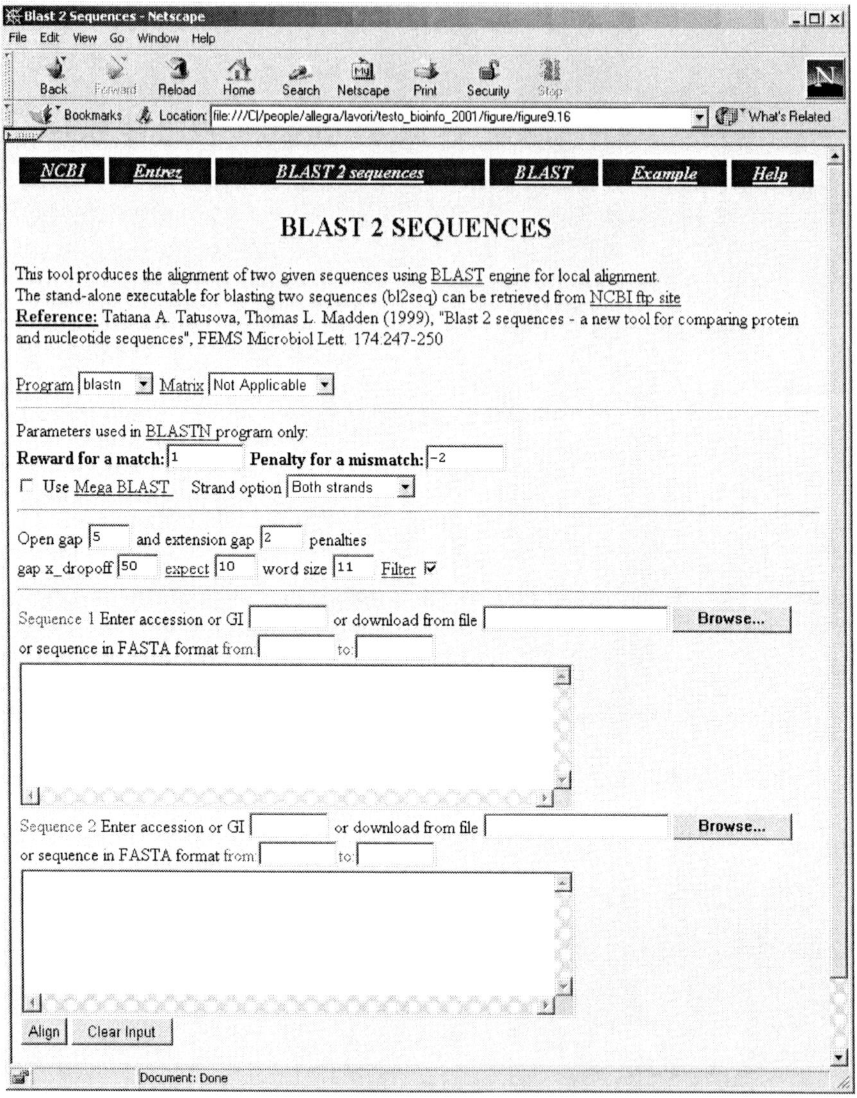

Figure 9.16. BLAST 2 www page.

Alignment view

The default alignment view is a pair-wise alignment of the query sequence and the database matches. However other options are provided by a pulldown menu.

Get results by e-mail

By typing or pasting your e-mail address, you can receive a copy of the BLAST results of your search. The graphical overview is not available by e-mail.

9.2.4 BLAST output

When you successfully run BLAST you obtain a '*formatting* BLAST' page where you can find your request ID and you may change the formatting parameters. To obtain your result, you must press FORMAT!. The partial results of BLAST for the sequence of Figure 9.13 is shown in Figure 9.15.

The graphical overview of the set of sequences aligned to the query sequence allows you to have instantly a general idea of the results of your search.

'Descriptions' are lines of text summarizing the highest scoring alignments. The first column of a description line is the protein identifier and is linked to the corresponding NCBI database entry for that protein. The second column gives a short description of the protein, the third one reports the bit score of the alignment and the fourth the E value (see chapter 7). The higher the score the better the alignment, however the significance of the alignment is provided by the E value, i.e. the expected number of alignments with a score of S or better (see chapter 7) found by chance; namely generated using databases of random sequence of comparable length and composition. This number can be evaluated by looking at alignment scores generated using databases of random sequences of comparable length and composition. As the score increases, the E value exponentially decreases. The description lines are sorted by increasing E values.

Scrolling down the page you can see the alignments (only the second highest scoring alignment is reported in Figure 9.15). They can be listed in many different formats (see BLAST format options above). The default is the pair-wise alignment between the query sequence and the database match. Identical residues are listed between the two sequences, conserved residues are represented by '+' signs, gaps by dashes in one of the two sequences aligned and, due to filtering, low complexity subsequences by Xs in the query sequence.

9.2.5 Alignment of two sequences

As discussed at the beginning of the BLAST section, this program can be used also for the alignment of two given sequences. The tool for pair-wise sequence alignment can be accessed from the 'BLAST 2 Sequences' link of the BLAST home page. The BLAST 2 web page is shown in Figure 9.16. As in the case of BLAST for database searches, you must choose a program in a pull down menu (blastp for protein sequences comparison). After the parameter settings (gap opening and extension penalties, expectation value etc.), the first and the second sequence to be aligned have to be typed or pasted in the 'Sequence 1' and 'Sequence2' text windows, respectively. The sequences accession numbers or GI can be used instead. Then the 'align' button must be pressed to run the program. The typical output web page of a BLAST 2 run is shown in Figure 9.17.

There are many other servers that provide a pair-wise alignment tool. The Web addresses of some of them can be found in Table 9.9 and in a number of Web sites for sequence analysis.

9.2.6 PSI-BLAST

The PSI-BLAST (Position-specific iterative BLAST) (Altschul *et al.*, 1997) program allows highly sensitive database searches, when looking for very distant relatives. It is based on an initial BLAST search and uses, to improve its sensitivity, a *profile* (see Position-Specific Scoring Matrices in chapter 7). After an initial BLAST search, a profile is built on a multiple alignment of the highest scoring hits. In this profile the program gives high scores to highly conserved positions and low scores to weakly conserved positions of the multiple alignment. This first step profile is used to perform a second BLAST search, and the

---

Table 9.9. Web site addresses for pairwise alignments

GAP, SIM	http://genome.cs.mtu.edu/align/align.html
SIM at ExPASy	http://www.expasy.ch/tools/sim-prot.html
SIM, BLAST 2	http://dot.imgen.bcm.tmc.edu:9331/seq-search/alignment.html

```
BLAST 2 SEQUENCES RESULTS VERSION BLASTP 2.1.2 [Oct-19-2000]

Sequence 1 sp|P31946|143B_HUMAN 14-3-3 protein beta/alpha (Protein kinase
C inhibitor protein-1) (KCIP-1) (Protein 1054) - Homo sapiens (Human).
Length 244 (1 .. 244)

Sequence 2 sp|O49995|143B_TOBAC 14-3-3-like protein B - Nicotiana tabacum
(Common tobacco). Length 255 (1 .. 255)

NOTE:The statistics (bitscore and expect value) is calculated based on the
size of nr database

Score =  299 bits (757), Expect = 3e-80
Identities = 155/243 (63%), Positives = 189/243 (76%), Gaps = 5/243 (2%)


Query: 2    MDKSELVQKAKLAEQAERYDDMAAAMKAVTEQ-GHELSNEERNLLSVAYKNVVGARRS 58
            M + E V  AKLAEQAERY++M + M+ V+         EL+ EERNLLSVAYKNV+GARR+
Sbjct: 1    MAREENVYMAKLAEQAERYEEMVSFMEKVSTSLGTSEELTVEERNLLSVAYKNVIGARRA 60

Query: 59   SWRVISSIEQKTER-NEKKQQMGKEYREKIEAELQDICNDVLELLDKYLIPNATQPESK 116
            SWR+ISSIEQK E    NE    +  +EYR KIE+EL +IC+ +L+LLD  LIP+A+  +SK
Sbjct: 61   SWRIISSIEQKEESRGNEDHVKCIQEYRSKIESELSNICDGILKLLDSCLIPSASAGDSK
120

Query: 117  VFYLKMKGDYFRYLSEVASGDNKQTTVSNSQQAYQEAFEISKKEMQPTHPIRLGLALNFS
176
            VFYLKMKGDY RYL+E  +G  ++     ++   AY+ A  +I+   E+ PTHPIRLGLALNFS
Sbjct: 121  VFYLKMKGDYHRYLAEFKTGAERKEAAESTLSAYKAAQDIANAELAPTHPIRLGLALNFS
180

Query: 177  VFYYEILNSPEKACSLAKTAFDEAIAELDTLNEESYKDSTLIMQLLRDNLTLWTSENQGD
236
            VFYYEILNSP++AC+LAK AFDEAIAELDTL EESYKDSTLIMQLLRDNLTLWTS+ Q D
Sbjct: 181  VFYYEILNSPDRACNLAKQAFDEAIAELDTLGEESYKDSTLIMQLLRDNLTLWTSDMQDD
240

Query: 237  EGD 239
             D
Sbjct: 241  GAD 243
```

Figure 9.17. BLAST 2 output. The two sequences aligned (143B_HUMAN and
143B_TOBAC) show the 63% of identical residues, 76% of conserved residues and 2%
of gaps. Identical residues are listed in the alignment line between the two sequences in
upper case characters, conserved residues are represented by '+' signs and gaps by dashes
in one of the two sequences aligned. The significativity of the alignment is given by the
'Expect' value.

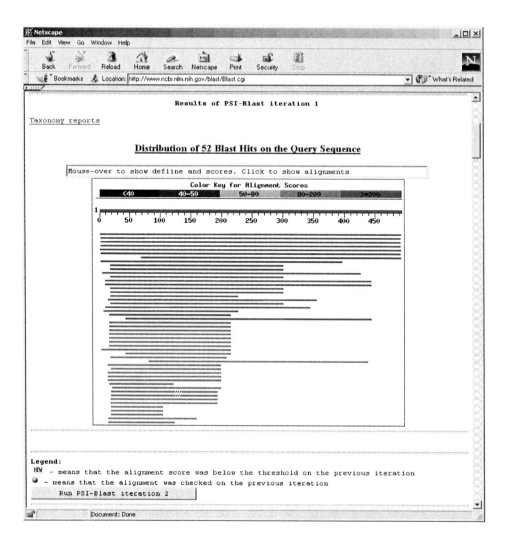

Figure 9.18a. PSI-BLAST output www page (part A).

results of this search are used to improve the sensitivity of the profile. Many iterations of this procedure can be performed if desired.

You can find the PSI-BLAST program in the NCBI BLAST home page (see Figure 9.11) by pressing the PSI- and PHI-BLAST link button. The PSI-BLAST www form is identical to the BLAST one (see Figure 9.12) except that some default optional parameters are turned off or on.

Figure 9.18b. PSI-BLAST output www page (part B).

In the 'Options for advanced blasting' section PSI-BLAST has no default filters turned on; in the 'Format' section the default number of descriptions and alignments is 500 and 250 respectively. The default 'Format for PSI-BLAST' option is clearly turned on.

You have to fill the form as described in the BLAST section.

The query sequence formats allowed, the set of databases you can choose and the discussion of the most important parameters are exactly the same than for BLAST.

One interesting option for advanced blasting that concerns only PSI-BLAST is indicated in the form page as <u>PSSM</u>. If you want to use a Position-Specific Scoring Matrix in other protein searches, you can save it in a file and then paste it in the PSSM text window.

How does one save a PSSM in a file? First of all you have to run a protein PSI-BLAST search. In the output page (see Figure 9.18) you must press the 'Run PSI-BLAST Iteration 2' button. Then go to the Format page, choose 'PSSM' from the 'Show' pull down menu and click 'Format'. This will display the text output with the ASCII-encoded PSSM that can be saved in a text file. The same procedure can be used to save in a file the PSSM of iteration n, where n is the number of iterations needed to obtain the 'complete' multiple alignment (i.e. the one including all the members of a family).

9.2.7 PSI-BLAST output

The PSI-BLAST output www page is very similar to the BLAST output page (sse Figure 9.18). The main difference consists of the 'Run PSI-Blast iteration 2' button just before AND after the list of sequences producing significant alignments. On clicking this button the programs builds a profile on a multiple alignment of the highest scoring (or with the lowest E values) sequences found with the first search. The number of sequences to be used to build the multiple alignment can be decided by fixing a threshold for the E value (i.e. the multiple alignment is constructed with all the sequences with an E value equal

Table 9.10. List of ClustalW www sites

ClustalW at EBI	http://www2.ebi.ac.uk/clustalw/
ClustalW at BCM	http://dot.imgen.bcm.tmc.edu:9331/multi-align/Options/clustalw.html
ClustalW at PBIL	http://pbil.univ-lyon1.fr/
ClustalW at Pasteur	http://bioweb.pasteur.fr/seqanal/alignment/intro-uk.html
ClustalW at DDBJ	www.ddbj.nig.ac.jp/searches-e.html

or lower then the threshold). Alternatively, the sequences can be selected manually. This profile is used for the second PSI-BLAST iteration when you click 'Format' in the Format page. The new results www page has a 'Run PSI-Blast iteration 3' button, and so on.

9.3 Multiple alignments

In this section the most widely used multiple alignment algorithms are presented. However a list of www sites where it is possible to find multiple sequence alignment methods is given in Table 9.13. As it has been said above (see chapter 7) the majority of methods performing multiple sequence alignments are based on *progressive alignments*. In the following sections the ClustalW (Higgins *et al.* 1996) and MultiAlign (Corpet, 1988) programs are described. These algorithms allow you to align any set of n ($n >1$) sequences. We note that the results obtained by ClustalW usually differ from those obtained by MultiAlign. This is true in general: multiple sequence alignments obtained with different methods differ not only because they are based on different principles (optimal global multiple alignments, progressive global multiple alignments, block-based global multiple alignments or motif-based local multiple alignments, see chapter 7) but also because the scoring system of each algorithm usually differently weights the optional parameters. This does not imply that one method is better than the others: the nature of the input sequences plays a role in the success of an algorithm and in general the user should use more than one method and revise and compare by hand the final alignments obtained.

Multiple alignment databases are essentially databases of protein sequence families and/or domains. Therefore the reader may refer to section 9.5.1, part (a) where such databases are described.

9.3.1 CLUSTALW

In this section the use of the ClustalW program that can be found at EBI (see Table 9.10) is described. For other ClustalW web sites you can refer to Table 9.10. The ClustalW www page at EBI is reported in Figure 9.19.

The search can be run interactively. However, if you want to receive the results by e-mail you can tape or paste your email address in the 'YOUR EMAIL' box.

The sequences to be aligned can be entered in the sequence input box at the bottom of the www page in a suitable format. The program supports, among the others, the FASTA format (Figure 9.8), the SWISSPROT format and the PIR format. No spaces or empty lines must be left from the beginning to the end of the input.

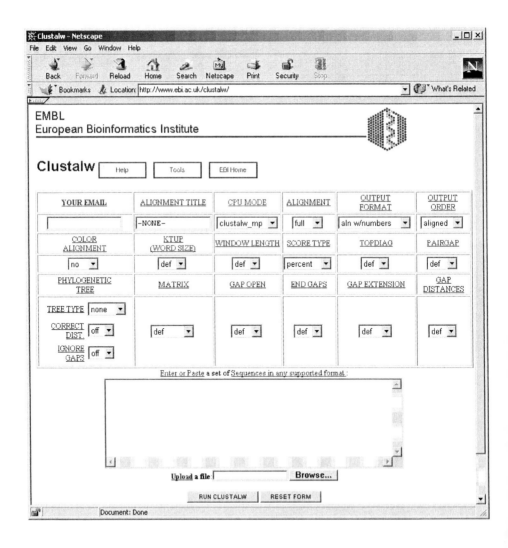

Figure 9.19. The ClustalW www page at EBI.

```
CLUSTAL W (1.81) multiple sequence alignment


sp|Q9CQV8|143B_MOUSE    TMDKSELVQKAKLAEQAERYDDMAAAMKAVTEQGHELSNEERNLLSVAYKNVVGARRSSW 60
sp|P35213|143B_RAT      TMDKSELVQKAKLAEQAERYDDMAAAMKAVTEQGHELSNEERNLLSVAYKNVVGARRSSW 60
sp|P29358|143B_BOVIN    TMDKSELVQKAKLAEQAERYDDMAAAMKAVTEQGHELSNEERNLLSVAYKNVVGARRSSW 60
sp|P31946|143B_HUMAN    TMDKSELVQKAKLAEQAERYDDMAAAMKAVTEQGHELSNEERNLLSVAYKNVVGARRSSW 60
                        ***********************************************************

sp|Q9CQV8|143B_MOUSE    RVISSIEQKTERNEKKQQMGKEYREKIEAELQDICNDVLELLDKYLILNATQAESKVFYL 120
sp|P35213|143B_RAT      RVISSIEQKTERNEKKQQMGKEYREKIEAELQDICSDVLELLDKYLILNATHAESKVFYL 120
sp|P29358|143B_bOVIN    RVISSIEQKTERNEKKQQMGKEYREKIEAELQDICNDVLQLLDKYLIPNATQPESKVFYL 120
sp|P31946|143B_HUMAN    RVISSIEQKTERNEKKQQMGKEYREKIEAELQDICNDVLELLDKYLIPNATQPESKVFYL 120
                        ***********************************  .***:******* ***:.******

sp|Q9CQV8|143B_MOUSE    KMKGDYFRYLSEVASGENKQTTVSNSQQAYQEAFEISKKEMQPTHPIRLGLALNFSVFYY 180
sp|P35213|143B_RAT      KMKGDYFRYLSEVASGDNKQTTVSNSQQAYQEAFEISKKEMQPTHPIRLGLALNFSVFYY 180
sp|P29358|143B_BOVIN    KMKGDYFRYLSEVASGDNKQTTVSNSQQAYQEAFEISKKEMQPTHPIRLGLALNFSVFYY 180
sp|P31946|143B_HUMAN    KMKGDYFRYLSEVASGDNKQTTVSNSQQAYQEAFEISKKEMQPTHPIRLGLALNFSVFYY 180
                        ****************:*******************************************

sp|Q9CQV8|143B_MOUSE    EILNSPEKACSLAKTAFDEAIAELDTLNEESYKDSTLIMQLLRDNLTLWTSENQGDEGDA 240
sp|P35213|143B_RAT      EILNSPEKACSLAKTAFDEAIAELDTLNEESYKDSTLIMQLLRDNLTLWTSENQGDEGDA 240
sp|P29358|143B_BOVIN    EILNSPEKACSLAKTAFDEAIAELDTLNEESYKDSTLIMQLLRDNLTLWTSENQGDEGDA 240
sp|P31946|143B_HUMAN    EILNSPEKACSLAKTAFDEAIAELDTLNEESYKDSTLIMQLLRDNLTLWTSENQGDEGDA 240
                        ********************************************.****************

sp|Q9CQV8|143B_MOUSE    GEGEN 245
sp|P35213|143B_RAT      GEGEN 245
sp|P29358|143B_BOVIN    GEGEN 245
sp|P31946|143B_HUMAN    GEGEN 245
                        *****
```

Figure 9.20. The ClustalW multiple alignment for the four sequences 143B_MOUSE, 143B_RAT, 143B_BOVIN and 143B_HUMAN. The symbol '*' under a column of a multiple alignment indicates 'identitical or conserved residues', the symbol ":" indicates conserved substitutions and "." indicates semi-conserved substitutions.

Many optional parameters can be specified from a series of pull down menus. Some of them are similar to that described for pair-wise alignment programs like BLAST (*ktup* (*word size*), *matrix*, *gap open*, *end gaps*, *gap extension*, *gap distances*). The others are well explained in the ClustalW Help that can be accessed, for each parameter, by clicking on the parameter itself (for example the help section for the CPU MODE parameter can be accessed by clicking on the CPU MODE link). If you do not have any particular requests, we suggest using the default parameters.

The ClustalW trivial multiple alignment for the four sequences 143B_MOUSE, 143B_RAT, 143B_BOVIN and 143B_HUMAN is shown in Figure 9.20.

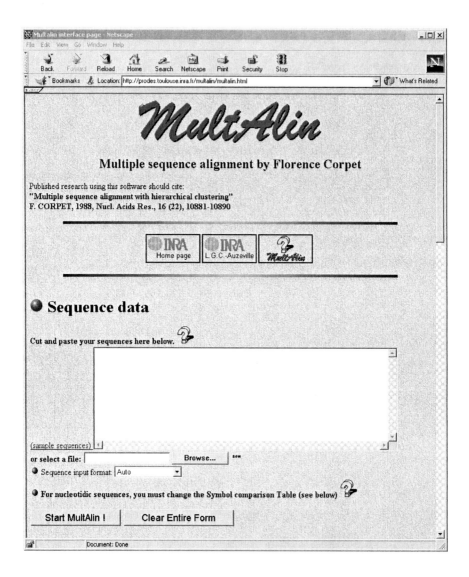

Figure 9.21. The MultiAlign Web page at the INRA Toulouse.

9.3.2 MultAlign

As for the ClustalW algorithm, this method is based on the idea of progressive alignments. It can be executed through the www sites given in Table 9.11. Here we describe the use of MultiAlign at the INRA Toulouse ('Multiple sequence alignment by F. Corpet') (Figure 9.21) The sequences to be aligned must be pasted in the text window in a

Table 9.11. List of MultiAlign www pages

MultiAlign at INRA Toulouse	http://prodes.toulouse.inra.fr/multalin/multalin.html
MultiAlign at PBIL	http://npsa-pbil.ibcp.fr/cgi-bin/
	npsa_automat.pl?page=npsa_multalin.html
MultiAlign at ETH Zurich	http://cbrg.inf.ethz.ch/Server/MultAlign.html

suitable format (fasta or SWISSPROT for protein sequences) that has to be specified using the 'Sequence input format' pulldown menu. Instead of pasting your sequences, you can give the name of your sequence file, or select it with the 'Browse' button. Then the program parameters can be chosen from a series of pulldown menus. As in the case of ClustalW the optional parameters include the output format (coloured gif image (default), plain text or coloured HTML text) the most suitable matrix to use (depending on the evolutionary distances between the input sequences) and the gap opening and gap extension penalties. Then you must click on the 'start MultiAlign!' button (at the top or at bottom of the page).

In Figure 9.22 the multiple alignment in plain text for the four sequences 143B_MOUSE, 143B_RAT, 143B_BOVIN and 143B_HUMAN is shown.

In this very simple example (the multiple alignment of the 14.3.3 sequences from four different organisms: mouse, rat, bovin and human), it is clear that the results of ClustalW and MultiAlign methods are identical (compare Figure 9.20 and Figure 9.22). This happens because the four sequences aligned are so similar that the differences between the two algorithm do not influence the final multiple alignments. However, when comparing more distantly related sequences the resulting alignments could be (very or slightly) different and you must take into account what has been said at the beginning of this section (9.3).

9.3.3 Editing a multiple alignment

Programs for multiple alignments usually allow saving of results as simple text. In general this is not a good format for publication and it

Table 9.12. List of Web sites for editing and/or visualizing a multiple alignment

ALSCRIPT	http://www.compbio.dundee.ac.uk/Software/Alscript/alscript.html
Jalview	http://www2.ebi.ac.uk/~michele/jalview/contents.html
seaview	http://pbil.univ-lyon1.fr/software/seaview.html
CINEMA	http://www.bioinf.man.ac.uk/dbbrowser/CINEMA2.1/
BOXSHADE	http://www.ch.embnet.org/software/BOX_form.html

is not very helpful to interpret a multiple alignment. Moreover, results from automatic procedures almost invariably require manual editing. This often presents problems, as there is currently no standard format for output, storage and distribution of alignments.

Therefore graphical interfaces have been developed that allow the editing of multiple alignments and to colour or shade residues, annotate alignments, manually add or remove gaps, modify misaligned positions, select particular domains and save or print alignments in a format suitable for publication. Usually each alignment editor can provide some of the functionalities listed above.

Here we give a short description of some of the most commonly used alignment presentation methods. Unfortunately not all of them can be run interactively. However they can be downloaded from the Web sites listed in Table 9.12.

9.3.3.1 ALSCRIPT

ALSCRIPT (Barton, 1993) is NOT an alignment editor. It is a Unix and PC-based program that takes as input a multiple alignment in a suitable format (among the others, CLUSTAL - i.e. the ClustalW output - format is accepted) and produces a Postscript file that may be printed on a Postscript laser printer, or viewed using a Postscript previewer. Therefore with the ALSCRIPT program you cannot manipulate a multiple alignment (in the sense of extracting sub-alignments, inserting gaps, correcting manually misaligned positions etc.) but you may obtain a picture of publication quality.

With ALSCRIPT, individual residues, or regions of a multiple alignment can be grey (at any level of grey from black to white) shaded. The shading may be also residue-specific ('all Arg residues between

```
MSF:    245    Check:    0              . .
 Name:  sp|Q9CQV8|143B Len:    245  Check: 2411  Weight:  1.22
 Name:  sp|P35213|143B Len:    245  Check: 2088  Weight:  1.22
 Name:  sp|P29358|143B Len:    245  Check: 3948  Weight:  0.78
 Name:  sp|P31946|143B Len:    245  Check: 3432  Weight:  0.78
 Name:  Consensus      Len:    245  Check:  167  Weight:  0.00

//


                         1                                                  50
sp|Q9CQV8|143B_MOUSE  TMDKSELVQK AKLAEQAERY DDMAAAMKAV TEQGHELSNE ERNLLSVAYK
  sp|P35213|143B_RAT  TMDKSELVQK AKLAEQAERY DDMAAAMKAV TEQGHELSNE ERNLLSVAYK
sp|P29358|143B_BOVIN  TMDKSELVQK AKLAEQAERY DDMAAAMKAV TEQGHELSNE ERNLLSVAYK
sp|P31946|143B_HUMAN  TMDKSELVQK AKLAEQAERY DDMAAAMKAV TEQGHELSNE ERNLLSVAYK
           Consensus  TMDKSELVQK AKLAEQAERY DDMAAAMKAV TEQGHELSNE ERNLLSVAYK

                         51                                                100
sp|Q9CQV8|143B_MOUSE  NVVGARRSSW RVISSIEQKT ERNEKKQQMG KEYREKIEAE LQDICNDVLE
  sp|P35213|143B_RAT  NVVGARRSSW RVISSIEQKT ERNEKKQQMG KEYREKIEAE LQDICSDVLE
sp|P29358|143B_BOVIN  NVVGARRSSW RVISSIEQKT ERNEKKQQMG KEYREKIEAE LQDICNDVLQ
sp|P31946|143B_HUMAN  NVVGARRSSW RVISSIEQKT ERNEKKQQMG KEYREKIEAE LQDICNDVLE
           Consensus  NVVGARRSSW RVISSIEQKT ERNEKKQQMG KEYREKIEAE LQDICnDVL#

                         101                                               150
sp|Q9CQV8|143B_MOUSE  LLDKYLILNA TQAESKVFYL KMKGDYFRYL SEVASGENKQ TTVSNSQQAY
  sp|P35213|143B_RAT  LLDKYLILNA THAESKVFYL KMKGDYFRYL SEVASGDNKQ TTVSNSQQAY
sp|P29358|143B_BOVIN  LLDKYLIPNA TQPESKVFYL KMKGDYFRYL SEVASGDNKQ TTVSNSQQAY
sp|P31946|143B_HUMAN  LLDKYLIPNA TQPESKVFYL KMKGDYFRYL SEVASGDNKQ TTVSNSQQAY
           Consensus  LLDKYLIlNA TqaESKVFYL KMKGDYFRYL SEVASG#NKQ TTVSNSQQAY

                         151                                               200
sp|Q9CQV8|143B_MOUSE  QEAFEISKKE MQPTHPIRLG LALNFSVFYY EILNSPEKAC SLAKTAFDEA
  sp|P35213|143B_RAT  QEAFEISKKE MQPTHPIRLG LALNFSVFYY EILNSPEKAC SLAKTAFDEA
sp|P29358|143B_BOVIN  QEAFEISKKE MQPTHPIRLG LALNFSVFYY EILNSPEKAC SLAKTAFDEA
sp|P31946|143B_HUMAN  QEAFEISKKE MQPTHPIRLG LALNFSVFYY EILNSPEKAC SLAKTAFDEA
           Consensus  QEAFEISKKE MQPTHPIRLG LALNFSVFYY EILNSPEKAC SLAKTAFDEA

                         201                                               245
sp|Q9CQV8|143B_MOUSE  IAELDTLNEE SYKDSTLIMQ LLRDNLTLWT SENQGDEGDA GEGEN
  sp|P35213|143B_RAT  IAELDTLNEE SYKDSTLIMQ LLRDNLTLWT SENQGDEGDA GEGEN
sp|P29358|143B_BOVIN  IAELDTLNEE SYKDSTLIMQ LLRDNLTLWT SENQGDEGDA GEGEN
sp|P31946|143B_HUMAN  IAELDTLNEE SYKDSTLIMQ LLRDNLTLWT SENQGDEGDA GEGEN
           Consensus  IAELDTLNEE SYKDSTLIMQ LLRDNLTLWT SENQGDEGDA GEGEN
```

Figure 9.22. Tthe multiple alignment in plain text for the four sequences 143B_MOUSE, 143B_RAT, 143B_BOVIN and 143B_HUMAN. In the consensus line at the bottom of the multiple alignment the residues that are identical in all the aligned sequences are in upper-case, the non-conserved residues are in lower-case and the symbol '#' indicates conserved substitutions.

Table 9.13. List of Web sites for multiple alignments

Optimal global multiple alignment
MSA at IBC   http://stateslab.bioinformatics.med.umich.edu/ibc/msa.html

Progressive global multiple alignment
MAP at BCM   http://dot.imgen.bcm.tmc.edu:9331/multi-align/multi-align.html
MAP   http://genome.cs.mtu.edu/map/map.html

Block-based global multiple alignment
DIALIGN 2   http://bibiserv.techfak.uni-bielefeld.de/dialign/ http://
                   www.expasy.ch/tools/sim-prot.html http://bioweb.pasteur.fr/
                   seqanal/interfaces/dialign2-simple.html
DCA   http://bibiserv.techfak.uni-bielefeld.de/cgi-bin/dca http://
                   bioweb.pasteur.fr/seqanal/interfaces/dca.html

Motif-based local multiple alignment
MATCH-BOX   http://www.fundp.ac.be/sciences/biologie/bms/
                   matchbox_submit.shtml

position 50 and 85'). Strings of text may be inserted at any position of the multiple alignment. Any Postscript font at any size and colour may be used on characters. Also, character backgrounds may be independently coloured. Rectangular boxes may be drawn around any part of the alignment and horizontal or vertical lines may be drawn to the left, right, top or bottom of any residue position or group of positions.

The available versions of the program can be downloaded from the Web site indicated in Table 9.12.

9.3.3.2 CINEMA and JALVIEW

With respect to ALSCRIPT (and many other alignment editor programs which are not mentioned here) CINEMA and JALVIEW have the advantage that they can be run interactively from a Web browser. Morover the user can directly use sequences, sequence databases, alignments or alignment databases available on the Web.

CINEMA (Colour INteractive Editor for Multiple Alignments) (Attwood et al., 1997; Parry-Smith et al., 1998) is a program for construction, modification, visualization and manipulation of alignments. It is a true multiple alignment editor. The input can be loaded from local sequence or alignment databases, pasted by the user in the suitable text box or loaded from a file.

The program allows the insertion/deletion of gaps (also simultaneously into set of sequences), the removal or re-ordering of sequences and the modification of characters font, size and colour.

The user can save the output file to a temporary directory for future program input. Alternatively, text, Postscript or GIF (for display within the browser) formats can easily be obtained. CINEMA is accessible via UCL's Bioinformatics server (see Table 9.12).

JALVIEW can be both downloaded or run directly from a WWW browser. As for CINEMA, JALVIEW also is a true multiple alignment editor. The program allows many input formats (CLUSTAL, FASTA, PIR, PFam and others) that can be loaded from a file or pasted in the input text box. You can obtain a Postscript output (coloured version of your alignment) by email or, alternatively an alignment output via a text box.

With this program it is possible to select, copy, move or delete sequences, select residues or groups of residues, insert or delete gaps, select columns of the alignment and perform many other manual manipulations of the input. All the editing options such as modification of character font, size and colour are also present.

9.3.3.3 BOXSHADE

BOXSHADE is a program for pretty-printing multiple alignment output. The program itself does not do any alignment, you have to use a multiple alignment program like ClustalW and use the output of these programs as input for BOXSHADE. Ports to all major platforms are available, as described on the web site at http://www.isrec.isb-sib.ch/software/BOX_faq.html

9.4 Hidden Markov Models (HMMs)

A HMM is a statistical model similar to a profile (Gribskov *et al.*, 1987) that can be used as a means of sequence modelling, multiple alignment and profiling (Baldi *et al.*, 1994, Eddy *et al.*, 1995, Eddy,

1995). A HMM can be estimated from aligned sequences but also from unaligned sequences, in particular it represents a highly effective means of modelling a family of unaligned sequences or a common motif within a set of unaligned sequences. Hidden Markov Models can be used to obtain multiple alignments of homologous sequences but also as a database searching tool.

The mathematical description of a HMM and its training procedure is relatively straighforward but its treatment is outside the scope of this chapter. However, the reader that is interested in the HMM statistical theory can look at the Hidden Markov Model Toolbox and Hidden Markov Models theory WWW sites indicated in Table 9.14 (first and second line, respectively) or refer to the wide literature on the subject. For a general introduction to HMMs see (Rabiner, 1989). There are several databases of protein families (see in section 9.5.1.3) and database searching tools (for example see the meta-MEME, Multiple Expectation-maximization for Motif Elicitation, software package Table 9.14; Grundy *et al.*, 1997) that are based on HMMs.

9.5 Motifs and patterns

Database similarity searches and/or multiple alignments algorithms in many cases may help the molecular biologist in the analysis of a newly-determined sequence or set of sequences. However, if the query sequence does not reveal a (global or local) similarity with another

Table 9.14. List of WWW sites for HMMs and programs for sequence analysis and alignments based on HMMs.

Hidden Markov Model (HMM) Toolbox	http://www.cs.berkeley.edu/~murphyk/Bayes/hmm.html
Manual for theory of Hidden Markov Models	http://jedlik.phy.bme.hu/~gerjanos/HMM/node2.html
HMMER2.2 - Profile hidden Markov models for biological sequence analysis	http://hmmer.wustl.edu/
SAM – Sequence alignment and modeling system	http://www.cse.ucsc.edu/research/compbio/sam.html
HMMER : using HMM (Hidden Markov Model)	http://bioweb.pasteur.fr/seqanal/motif/hmmer-uk.html
HMMER : hmmbuild - construct an HMM from a multiple sequence alignment	http://www-bioweb.pasteur.fr/seqanal/interfaces/hmmbuild.html
Multiple alignment and multiple sequence based searches	http://www.genetics.wustl.edu/eddy/publications/tigs-9808/
Meta-MEME - Motif-based hidden Markov modeling of biological sequences	http://metameme.sdsc.edu/

Table 9.15. The two-character codes which refers to the type of data contained in the lines of a PROSITE entry.

Line code	Content	Occurrence in an entry
ID	IDentification	Once; starts the entry
AC	ACcession number(s)	One per entry
DT	DaTe	One per entry
DE	short DEscription	One per entry
PA	PAttern	>=0 per entry
MA	MAtrix/profile	>=0 per entry
RU	RUle	>=0 per entry
NR	Numerical Results	>=0 per entry
CC	Comments	>=0 per entry
DR	Cross-references to SWISSPROT	>=0 per entry
3D	Cross-references to PDB	>=0 per entry
DO	pointer to the DOcumentation file	One per entry
//	Termination line	Once; ends the entry

sequence of known function or structure (for homology modelling, see Chapter 10), the results of the search will be not satisfactory. However, there are proteins that are related because of the common occurrence of a few critical residues: proteins sharing a functional site may display conserved residues on the surface of their three-dimensional structures (see chapter 10). These conserved residues are not necessarily contiguous in sequence, however, when taken together they can identify a sequence *pattern* or *motif* with fixed positions and alternating positions that can accomodate any amino acid. The presence of such a pattern or motif in a newly-determined sequence can be used to infer the function of the corresponding protein.

In this section some methods for *pattern matching* (i.e. the identification of patterns in sequences) are described. These methods rely on databases of patterns or motifs and on computer programs scanning a query sequence or set of sequences against such databases. In the following the most important databases of functional signatures or protein sequence domains are described and some computer programs searching patterns or motifs in a sequence or set of sequences are discussed.

The manual procedures and automatic methods for the exctraction or construction of a new pattern or motif from a set of proteins sharing a functional domain comprises a complex subject. However this interesting subject is outside the scope of this book and will not be discussed here.

9.5.1 Pattern and domain databases

9.5.1.1 PROSITE

PROSITE is one of the most widely used databases (Bairoch, 1991) of protein families and domains. The release 16.46, of 27-Sep-2001, contains approximately 1500 different entries consisting of *patterns*, *rules* and *profiles/matrices*. These are objects encoding signatures specific for about a thousand of protein families or domains. PROSITE entries are extremely well documentated. You can access PROSITE from the ExPASy home page (http://www.expasy.ch/) or using the web address indicated in Table 9.17.

The database, largely manually maintained, essentially consists of two text files: PROSITE.DAT and PROSITE.DOC. The file PROSITE.DAT contains a list of all the database entries structured so as to be usable by readers as well as by computer programs. Each

Table 9.16. The type of data that can be found in NR lines of a PROSITE entry

/RELEASE	SWISS-PROT release number and total number of sequence entries in that release.
/TOTAL	Total number of hits in SWISS-PROT.
/POSITIVE	Number of hits on proteins that are known to belong to the set in consideration.
/UNKNOWN	Number of hits on proteins that could possibly belong to the set in consideration.
/FALSE_POS	Number of false hits (on unrelated proteins).
/FALSE_NEG	Number of known missed hits.
/PARTIAL	Number of partial sequences which belong to the set in consideration, but which are not hit by the pattern or profile because they are partial (fragment) sequences.

```
ID   PPASE; PATTERN.
  AC   PS00387;
  DT   NOV-1990 (CREATED); NOV-1995 (DATA UPDATE); NOV-1995 (INFO UPDATE).
  DE   Inorganic pyrophosphatase signature.
  PA   D-[ SGN] -D-[ PE] -[ LIVM] -D-[ LIVMGC] .
  NR   /RELEASE=32,49340;
  NR   /TOTAL=16(16); /POSITIVE=11(11); /UNKNOWN=0(0); /FALSE_POS=5(5);
  NR   /FALSE_NEG=0; /PARTIAL=2;
  CC   /TAXO-RANGE=A? EP? ; /MAX-REPEAT=1;
  CC   /SITE=1,magnesium; /SITE=3,magnesium; /SITE=6,magnesium;
  DR   P21216, IPYR_ARATH, T; P37980, IPYR_BOVIN, T; P17288, IPYR_ECOLI, T;
  DR   P44529, IPYR_HAEIN, T; P13998, IPYR_KLULA, T; P19117, IPYR_SCHPO, T;
  DR   P37981, IPYR_THEAC, T; P19514, IPYR_THEP3, T; P38576, IPYR_THETH, T;
  DR   P00817, IPYR_YEAST, T; P28239, IPY2_YEAST, T;
  DR   P19371, IPYR_DESVH, P; P21616, IPYR_PHAAU, P;
  DR   P09167, AERA_AERHY, F; P12351, CYP1_YEAST, F; P24653, Y101_NPVOP, F;
  DR   P37904, YCEI_ECOLI, F; P39303, YJFU_ECOLI, F;
  3D   1PYP;
  DO   PDOC00325;
  //
```

Figure 9.23. Example of a pattern entry. Underlined items are linked to the corresponding web pages (SWISSPROT entries or documentation files).

entry is composed of different types of lines, each with their own format, beginning with a two-character code which refers to the type of data contained in the corresponding line. A short description of the two-character codes is given in Table 9.15.

An example of a pattern entry is shown in Figure 9.23. The ID, AC, DT, DE, DR, 3D, DO and '//' lines do not require a detailed explanation (which can be found, if needed, in the PROSITE user manual). There are protein families for which it is not possible to extract a pattern in the form of a single regular expression. In such cases it can be possible to construct a profile (Gribskov, 1987) with more expressive power than a pattern, that can identify members of a family better than a pattern. MA lines refer to profiles. For the sake of simplicity, they will not be described here in detail. For biologists that are interested, the structure of a PROSITE matrix/profile or rule entry can be found in the PROSITE user manual.

Here we want to discuss briefly the PA, the NR and the CC lines.

The PA lines contain the definition of a PROSITE pattern. The syntax adopted is as follows:

(a)   The standard IUPAC one-letter codes for the amino acids are used.

(b)   A specified residue allowed in a position is indicated with one character

(c)   Positions where any amino acid is accepted are indicated with the symbol 'x'

(d)   [...] denotes a set of allowed residues at a given position. For example [DE] stands for Asp or Glu.

(e)   {...} denotes a set if residues that are not accepted at a given position. For example: {AL} stands for any amino acid except Ala or Leu.

(f)   Each position in the motif is separated by a hyphen.

(g)   (n) indicates a repeat of n. For example: R(4) stands for a repeat of four Arg.

(h)   (n,m) indicates a repeat of between n and m inclusive. For example: x(3,5) stands for a repeat of three, four or five positions where any amino acid is accepted.

(i)   The '<' symbol starting the pattern and/or the '>' symbol ending the pattern stand for a pattern restricted to the sequence N- or C-terminal respectively.

(l)   A period ends the pattern.

The NR lines contain results deriving from a scan with the pattern on the SWISSPROT sequence database. In other words they provide information on the pattern reliability. The type of data that can be found in NR lines are shortly described in Table 9.16.

The CC lines provide comments such as the taxonomic range of a pattern or profile, the maximum number of times a given pattern or profile has been found in a single protein sequence or the position of an 'interesting' site in a pattern or a profile. The syntax adopted for these lines is described in the 'CC lines' part of the PROSITE user manual.

The two text files PROSITE.DAT and PROSITE.DOC can be obtained following the instructions of the 'How to obtain PROSITE' link in the PROSITE database home page.

Data contained in these text files can however be accessed directely on the web clicking one of the link buttons listed in the

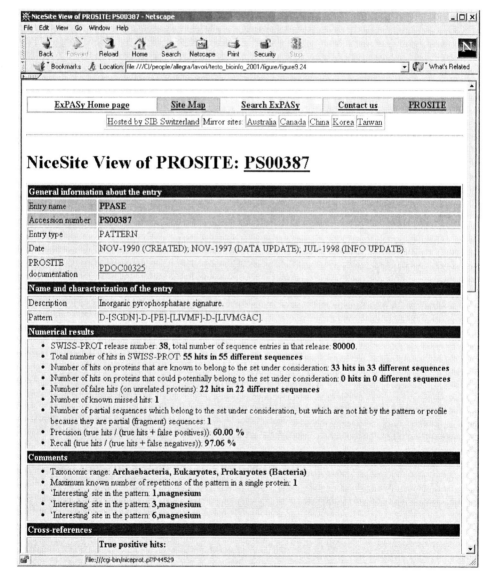

Figure 9.24. The 'Nice view' of the entry of Figure 9.23.

'accesss to PROSITE' section of the database home page. In this case you obtain the very well organized 'Nice view' of the PROSITE entries and documentation. The 'Nice view' of the entry of Figure 9.23 is reported in Figure 9.24.

The PROSITE.DOC text file mentioned above contains the full documentation for the PROSITE entries. The documentation for an entry provides a description of the family identified by the pattern and its biological role. Moreover it also indicates eventual biologically important sites and gives literature references. Links to experts for more information can be found in the documentation 'Nice view' Web page.

9.5.1.2 BLOCKS

BLOCKS (Henikoff *et al.*, 1991; Pietrokovski *et al.*, 1996) is a protein family database in which the same families of PROSITE are represented by means of ungapped local multiple alignments called 'blocks', instead of patterns or profiles. The construction of a 'block' is made using automated procedures by looking for the most highly conserved regions in groups of proteins documented in the PROSITE database. The database can be accessed thorough the BLOCK server (see section 9.5.2.2 and Figure 9.32) links: 'Get Blocks by key word' and 'Get Blocks by number'.

If you know the BLOCKS AC (accession number) then you can type it in the text box at the top of the 'Get Blocks by number' Web page. However this will not be the most common situation. In general you will have a key word or a set of key words or a sequence database (SWISSPROT, TREMBL, PIR etc.) ID or name for your protein sequence. In this case you must access the 'Get Blocks by key word' Web page and tape or paste the key word(s) or protein name in the text box at the top of the page. The page also provides an help for filling the text box in properly.

As an example, imagine that you want to identify ungapped segments corresponding to highly conserved regions of proteins involved in phosphorylation (e.g. phosphorylation sites). You can type 'phosphorylation' in the text box and press RETURN to obtain the result shown in Figure 9.25.

BLOCKS Database Keyword Search

Note that you can enter a new query in the search term box from this screen/page without having to go back.

The following 2 item(s) match your query '**phosphorylation**':

- IPB002114
    PTS_HPr_ser Serine phosphorylation site in HPr protein

- IPB001020
    PTS_HPr_his Histidine phosphorylation site in HPr protein

Figure 9.25. The result of the 'Get Blocks by key word' search performed with the keyword 'phosphorylation'.

Choosing, for instance, the 'Serine phosphorylation site in HPr protein', clicking on IPB002114, you obtain the BLOCKS entry page, containing two blocks i.e. two short multiply aligned ungapped segments corresponding to the most highly conserved regions in the HPr proteins:

Block IPB002114A

```
ID PTS_HPr_ser; BLOCK
AC IPB002114A; distance from previous block=(1,3)
DE Serine phosphorylation site in HPr protein
BL GAR; width=31; seqs=4; 99.5%=1253; strength=1194


PT1A_ECOLI|P32670 ( 3) LIVEFICELPNGVHARPASHVETLCNTFSSQ 100

YPDD_ECOLI|P77439 ( 2) LTIQFLCPLPNGLHARPAWELKEQCSQWQSE 92

PTHP_XYLFA|Q9PDH6 ( 2) LEHELIVTNKLGLHARATAKLVQTMSKFQSN 94

PTSO_SHEVI|Q9S0K8 ( 4) LERQVTICNKLGLHARAATKLAILASEFDAE 94

//

and
```

Block IPB002114B

```
ID PTS_HPr_ser; BLOCK
AC IPB002114B; distance from previous block=(11,574)
DE Serine phosphorylation site in HPr protein
BL LGD; width=44; seqs=4; 99.5%=1728; strength=1189
PT1A_ECOLI|P32670 ( 608)
GKWIGLCGELGAKGSVLPLLVGLGLDELSMSAPSIPAAKARMAQ 77

YPDD_ECOLI|P77439 ( 605)
GKWVGICGELGGESRYLPLLLGLGLDELSMSSPRIPAVKSQLRQ 79

PTHP_XYLFA|Q9PDH6 ( 44)
AKSIMGVMLLAASQGTVIRVRIDGEDEHTAMQALSELFENRFNE 100

PTSO_SHEVI|Q9S0K8 ( 46)
AASVLGLLMLETGMGKTITLLGKGQDADAALDAICALVDAKFDE 99

//
```

As in the PROSITE database, the annotation lines begin with a two-character code. ID and AC indicates the BLOCKS identification and the accession number, respectively. DE stands for (short) description while BL is PROTOMAT information (PROTOMAP is an automatic hierarchical classification of all SWISSPROT and TrEMBL proteins. See http://protomap.stanford.edu/ or http://protomap.cornell.edu/). The underlined sequence protein SWISSPROT name preceding each segment is a link to the corresponding entry. Segments are clustered if >=80% of aligned residues match between any pair of segments. For each segment, the SWISS-PROT name is followed by the position of the first residue in the segment.

Table 9.17. List of Web sites of patterns and domains databases

PROSITE	http://www.expasy.ch/prosite/
PFam home page (St. Louis)	http://pfam.wustl.edu/
PFam home page (France)	http://pfam.jouy.inra.fr/
PFam (Cambridge)	http://www.sanger.ac.uk/Software/Pfam/
PFam (Stockholm)	http://www.cgr.ki.se/Pfam/
BLOCKS	http://www.blocks.fhcrc.org/
PRINTS	http://www.bioinf.man.ac.uk/dbbrowser/PRINTS/
ProDom	http://protein.toulouse.inra.fr/prodom/doc/prodom.html

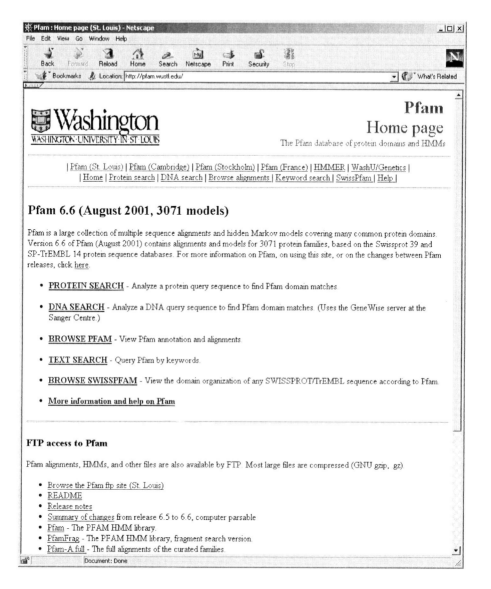

Figure 9.26. The PFam WWW home page.

9.5.1.3 PFam

PFam (Sonnhammer *et al.*, 1997; 1998) is a large collection of multiple sequence alignments and hidden Markov models of protein families and complete protein domains. The database Web sites provide, besides the database itself, tools for querying PFam by keywords (KEYWORD

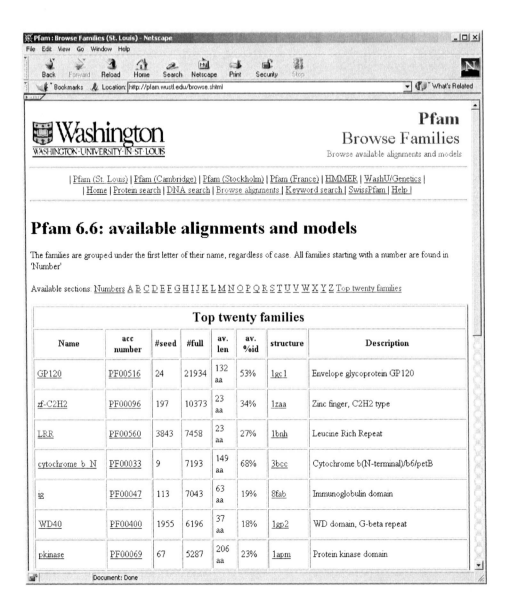

Figure 9.27. The top WWW page for browsing PFam. The (8x20) table contains the list of the top twenty families of the database. First column: PFam family name; 2nd column: PFam accession number; 3rd column: number of sequences of the seed alignment; 4th column: number of sequences of the full alignment; 5th column: average number of residues per sequence aligned; 6th column: average percent identity between the sequences aligned; 7th column: link to a representative structure of the family (if existing); 8th column: short description of the family.

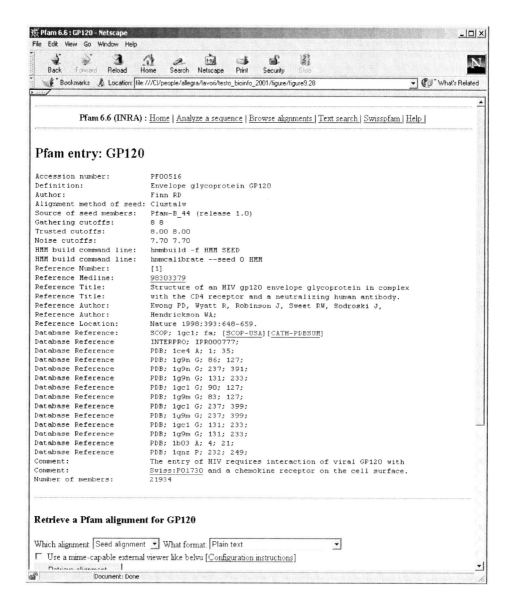

Figure 9.28. The GP120 PFam entry.

SEARCH), find PFam domain matches in a protein query sequence (PROTEIN SEARCH) and find PFam domain matches by organism (TAXONOMY SEARCH). Also a special database SwissPFam is provided which shows the domain organization (according to PFam) of the protein sequences in SWISSPROT and TREMBL.

Table 9.18. Short description of the most important fields of the GP120 PFam entry

Accession number	The Pfam-A accession numbers PFxxxxx are the identifiers for each Pfam families
Definition	This must be a one line description of the Pfam family
Author	Author of the entry
Alignment method of seed	The method used to align the seed members
Source of seed members	The source suggesting seed members belong to a family
Gathering cutoffs	Search threshold to build the full alignment
Trusted cutoffs	It refers to the bit scores of the lowest scoring match in the full alignment. The first number refers to the lowest whole sequence score in bits of a match in the full alignment, and the second number specifies the lowest per-domain score in bits of a match in the full alignment. These two scores may not refer to the same sequence.
Noise cutoffs	It refers to the bit scores of the highest scoring match not in the full alignment. The first number refers to the highest whole sequence score in bits of a match not in the full alignment, and the second number specifies the highest per-domain score in bits of a match not in the full alignment. These two scores may not refer to the same sequence.
HMM build command line	See the HMMER 2 user's manual for full instructions on building HMMs at: http://hmmer.wustl.edu/
Reference Number	Reference numbers are used to precede literature references, which have multiple line entries
Reference Medline	The number can be found as the UI number in pubmed at: http://www.ncbi.nlm.nih.gov/PubMed/
Reference Title	Title of paper
Reference Author	Author(s) of paper
Reference Location	Location of paper
Database Reference	Reference to external database
Comment	Comment lines provide annotation and other information

The PFam home page can be found both at the Washington university in St. Louis and at the INRA (France) (see Table 9.17). It is shown in Figure 9.26.

In the following we describe how to view PFam annotation and alignments. By clicking the 'BROWSE PFAM' button, you obtain the 'available alignments and models' WWW page (Figure 9.27). You

```
ENV_HV1A2/33-509 LWVTVYYGVPVWKEATTTLFCASDARAYDTEVHNVWATHACVPTDPNPQEVVLGNVTENFNMWK..NNMVEQMQEDIISLWDQ
ENV_HV1B1/34-511 LWVTVYYGVPVWKEATTTLFCASDAKAYDTEVHNVWATHACVPTDPNPQEVVLVNVTENFNMWK..NDMVEQMHEDIISLWDQ
ENV_HV1J3/33-523 LWVTVYYGVPVWKEAATTLFCASDAKAYDTEVHNVWATHACVPTDPNPQEVVLENVTEKFNMWK..NNMVEQMHEDIISLWDQ
ENV_HV1W1/33-510 LWVTVYYGVPVWKEATTTLFCASDAKAYSTEAHKVWATHACVPTNPNPQEVVLENVTENFNMWK..NNMVEQMHEDIISLWDQ
ENV_HV1BN/34-507 LWVTVYYGVPVWKEANTTLFCASDAKAYDTEIHNVWATHACVPTDPNPQELVMGNVTENFNMWK..NDMVEQMHEDIISLWDQ
ENV_HV1BH/33-519 LWVTVYYGVPVWKEATTTLFCASEAKAYKTEVHNVWAKHACVPTDPNPQEVLLENVTENFNMWK..NNMVEQMHEDIISLWDQ
ENV_HV1QY/33-509 LWVTVYYGVPVWKEATTTLFCASDARAYATEVHNVWATHACVPTDPNPQEVVLGNVTENFDMWK..NNMVEQMQEDIISLWDQ
ENV_HV1C4/35-522 LWVTVYYGVPVWKEATTTLFCASDAKAYDTEAHNVWATHACVPTNPNPQEVVLENVTENFNMWK..NNMVEQMHEDIISLWDQ
ENV_HV1ZH/33-511 LWVTVYYGVPVWKDAETTLFCASDAKAYDTEKHNVWATHACVPTDPNPQELSLGNVTEKFDMWK..NNMVEQMHEDVISLWDQ
ENV_HV1EL/33-508 LWVTVYYGVPVWKEATTTLFCASDAKSYETEAHNIWATHACVPTDPNPQEIALENVTENFNMWK..NNMVEQMHEDIISLWDQ
ENV_HV1Z8/33-518 LWVTVYYGVPVWKEATTTLFCASDAKSYEPEAHNIWATHACVPTDPNPREIEMENVTENFNMWK..NNMVEQMHEDIISLWDQ
ENV_HV1ND/33-501 LWVTVYYGVPIWKEATTTLFCASDAKAYKKEAHNIWATHACVPTDPNPQEIELENVTENFNMWK..NNMVEQMHEDIISLWDQ
ENV_HV1MA/33-513 LWVTVYYGVPVWKEATTTLFCASDAKSYETEVHNIWATHACVPTDPNPQEIELENVTEGFNMWK..NNMVEQMHEDIISLWDQ
ENV_SIVCZ/33-496 LWVTVYYGVPVWHDADPVLFCASDAKAHSTEAHNIWATQACVPTDPSPQEVFLPNVIESFNMWK..NNMVDQMHEDIISLWDQ
ENV_HV2BE/24-510 QYVTVFYGIPAWKNASIPLFCATKNR.......DTWGTIQCLPDNDDYQEIILN.VTEAFDAWN..NTVTEQAVEDVWHLFET
ENV_HV2CA/25-512 QYVTVFYGVPAWKNASIPLFCATKNR.......DTWGTIQCLPDNDDYQEIPLN.VTEAFDAWD..NTITEQAIEDVWNLFET
ENV_HV2D1/24-501 QYVTVFYGIPAWRNASIPLFCATKNR.......DTWGTIQCLPDNDDYQEITLN.VTEAFDAWD..NTVTEQAIEDVWRLFET
ENV_HV2G1/23-502 QYVTVFYGVPAWRNASIPLFCATKNR.......DTWGTIQCKPDNDDYQEITLN.VTEAFDAWD..NTVTEQAVEDVWSLFET
ENV_HV2NZ/24-502 QFVTVFYGIPAWRNASIPLFCATKNR.......DTWGTIQCLPDNDDYQEITLN.VTEAFDAWN..NTVTEQAVEDVWNLFET
ENV_SIVM1/24-528 QYVTVFYGVPAWRNATIPLFCATKNR.......DTWGTTQCLPDNDDYSELALN.VTESFDAWE..NTVTEQAIEDVWQLFET
ENV_HV2D2/24-513 QYVTVFYGIPAWRNATVPLICATTNR.......DTWGTVQCLPDNGDYTEIRLN.ITEAFDAWD..NTVTQQAVDDVWRLFET
ENV_SIVA1/24-538 QWITVFYGVPVWKNSSVQAFCMTPTT.......RLWATTNCIPDDHDYTEVPLN.ITEPFEAWADRNPLVAQAGSNIHLLFEQ
ENV_SIVAI/22-522 LYVTVFYGIPVWKNSTVQAFCMTPNT.......NMWATTNCIPDDHDNTEVPLN.ITEAFEAWD..NPLVKQAESNIHLLFEQ
ENV_SIVGB/47-569 QYVTVFYGVPVWKEAKTHLICATDNS.......SLWVTTNCIPSLPDYDEVEIPDIKENFTGLIRENQIVYQAWHAMGSMLDT
```

Figure 9.29. The partial seed multiple alignment of the GP120 PFam entry. The underlined items are the SWISSPROT sequence identifiers and are linked to the corresponding SWISSPROT entries.

can browse the database starting from the top twenty families or by selecting the name of a family (the family names are grouped in alphabetical order with an index). The table of Figure 9.27 contains the list of the top twenty families of the database. The family names are listed in the first column and clicking on one of them you can have access to the corresponding entry. For example, if you click on the first family name (GP120) you obtain the entry shown in Figure 9.28. The GP120 family alignment can be retrieved by clicking the 'retrieve alignment' botton in the 'Retrieve a PFam alignment for GP120' section of the entry.

It should be noted that both in the third and fourth column of the table of figure 9.27 and in the first pulldown menu of the 'Retrieve a PFam alignment for GP120' section of the GP120 entry (Figure 9.28) two types of alignment ('seed alignment' and 'full alignment') are mentioned.

Seed alignments are of representative, nonredundant sets of sequences whereas full alignments are HMM-generated automatic alignments of every homologous domain in the current release of SWISSPROT that is detected by an HMM built from the seed alignment. Seed alignments are manually checked and are stable from release to release.

Table 9.19. Short description of instructions to fill the box text of PRINTS ways of access.

Way of access	How to fill the corresponding text box
Search by accession number	type or paste an accession number (e.g. pr000189 for the kringle domain)
Search by PRINTS code	type or paste a PRINTS code (e.g. kringle for kringle domains)
Search by database code	type or paste a database code (e.g. 143B_HUMAN retrieves entries containing this SWISSPROT code)
Search by text	type or paste a text string (e.g. rhodopsin retrieves entries in which 'rhodopsin' appears)
Search by sequence	type or paste a sequence fragment (e.g dryf displays entries containing the sequence Asp-Arg-Tyr-Phe)
Search database titles	type or paste a text string (e.g. rhodopsin lists entries in which 'rhodopsin' appears in the source database title lines)
Search by number of motifs	type or paste a number (e.g. 9 lists all fingerprints containing 9 motifs)
Search by author	type or paste a author name (e.g. boguski lists all entries in which the name Boguski appears)
Search by query language	use the syntax /qualifier function query (e.g. /full code kringle retrieves the full entry for kringle domains)

A typical PFam entry (see Figure 9.28) has a format that very easy to read and understand. However, Table 9.18 provides a short description of the most important fields (such as 'Accession number', 'Definition', 'Author', etc.) of the entry. The seed multiple alignment of the GP120 PFam entry is partially reported in Figure 9.29.

9.5.1.4 PRINTS

The PRINTS database (Attwood *et al.*, 1994a; Attwood *et al.*, 1994b, Attwood *et al.*, 2002) is a collection of fingerprints. A fingerprint consists of a set of motifs characterizing a protein family. The motifs of a fingerprint are local ungapped alignments that usually are separated along the sequence but that could be contiguous in 3D-space.

The first step for the construction of a fingerprint is made manually making a multiple alignment of some members of a family.

Tab 9.20. List of WWW servers sites for scanning a sequence against databases of domains, motifs, paterns or profiles.

ProfileScan	http://www.isrec.isb-sib.ch/software/PFSCAN_form.html
Blocks	http://www.blocks.fhcrc.org/
SMART	http://coot.embl-heidelberg.de/SMART

Then the fingerprint is automatically scanned against a sequence database (SWISS-PROT/TrEMBL) until no new family members are found. Compared with a single motif, a fingerprint is more powerful in identifying a newly determined sequence as a member of a known family. As for PROSITE, the PRINTS entries are very well documented.

The PRINTS database can be accessed by accession number, by PRINTS code, by database code, by text, by sequence, by database title, by number of motifs, by author and by query language. The links to these different way of access can be found in the database Web home page (see Table 9.17) in the 'Direct PRINTS access' part. Each one of these links produces, when clicking on it, a standard Web page with a text box and the corresponding instructions to fill it. A short description of such instructions is given in Table 9.19. The multiple alignments of a PRINTS entry are visualized with the CINEMA program (see section 9.3.3.2).

The website of PRINTS also provides some useful tools for sequence analysis, such as FingerPRINTScan for scanning a query sequence against the database.

9.5.2 Servers for patterns and domains databases scanning

9.5.2.1 ProfileScan

The ProfileScan Server (see Table 9.20) uses a program called **pfscan** to compare a protein sequence to current releases of PROSITE profiles and patterns database and/or the PFam collection of Hidden Markov Models. The server home page is shown in Figure 9.30. The use of the

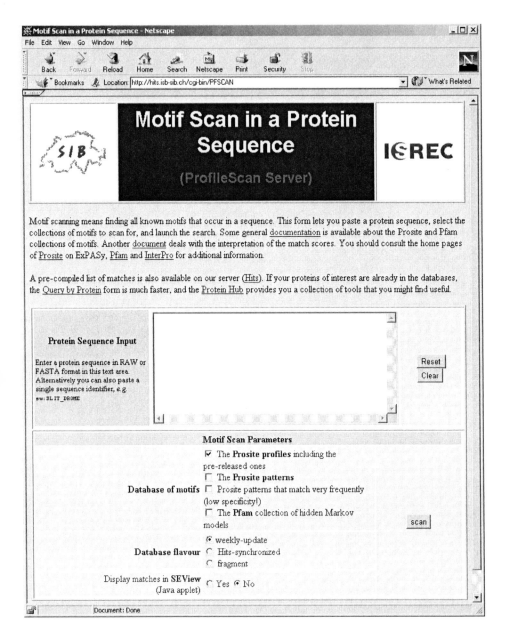

Figure 9.30. The ProfileScan server WWW home page.

pfscan program is straightforward. You have to paste a protein sequence in a suitable format (the fasta format is accepted) in the text window shown in the server website. Then you must choose the database of

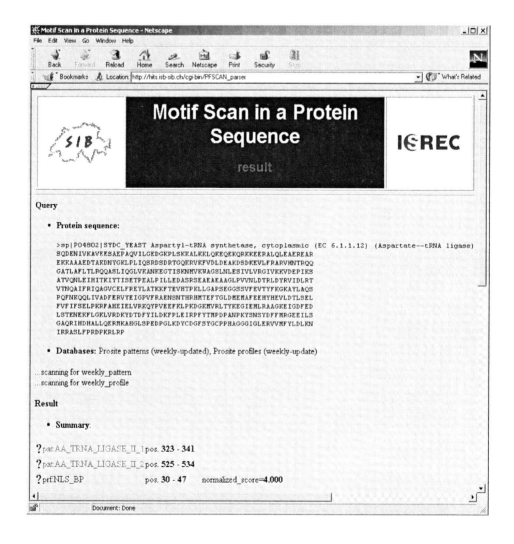

Figure 9.31a. The pfscan output for the sequence of the aspartyl-tRNA synthetase SYDC_YEAST using the collection af PROSITE patterns as a database.

motifs to be scanned with the query sequence and press the 'scan' button. The program run is quite fast and the output for the sequence of the aspartyl-tRNA synthetase SYDC_YEAST using the collection af PROSITE patterns as a database is reported in Figure 9.31.

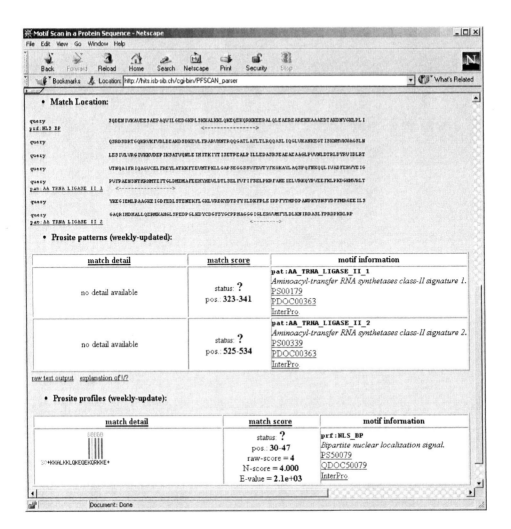

Figure 9.31b. The pfscan output for the sequence of the aspartyl-tRNA synthetase SYDC_YEAST using the collection af PROSITE patterns as a database (continued).

9.5.2.2 BLOCKS server

The BLOCKS server (see Table 9.20) (Henikoff *et al.*, 1998; Pietrokovski *et al.*, 1998) provides, besides the BLOCK database links ('Get Blocks by key word' and 'Get Blocks by number'), some tools for biological sequence analysis. The most important programs are listed in Table 9.21.

The Blocks searcher is a program that scans the Blocks database

Table 9.21. List of tools available at the Blocks server website

Program	Function
Block Searcher	to search a sequence vs Blocks
Reverse PSI-BLAST Searcher	to search a sequence vs Blocks using NCBI's RPS-BLAST program
Impala Searcher	to search a sequence vs Blocks using NCBI's IMPALA program
Block Maker	to create Blocks
Multiple Alignment Processor	to excise Blocks from multiple alignments
LAMA Searcher	to search Blocks vs Blocks
COBBLER	to search embedded Blocks vs sequence databases
CODEHOP	to design PCR primers from Blocks
Biassed Block Checker	to identify biassed blocks

and the PRINTS database with a query sequence. You can go to the Block Searcher sequence entry page (see Figure 9.32) by clicking the proper item in the Blocks Server website. The program can be run interactively. If you want the results sent by e-mail, enter your e-mail address in the suitable text box. Choose in the pulldown menu the database you want to scan and, if necessary, set the few optional parameters following the parameters links for help. Finally you must enter the query sequence (in fasta format) in the text window and press the 'perform search' botton at the bottom of the WWW page.

The results Web page begins with an introduction that contains explanations about the arrangement and meaning of the search results lower on the page. The section labelled "Hits", begins with the name and size of the query, the number of blocks searched, the number of alignments done and the cutoff E-values used. Next there is a list of hits ('Family') with a link to the database entry. The last part of the result page contains pairwise alignments of the hits with the query.

9.5.2.3 SMART server

SMART (see Table 9.20) (a Simple Modular Architecture Research Tool) (Shultz et al., 1998; Ponting et al., 1999) is a WWW tool and database that allows the identification of domains in a query sequence

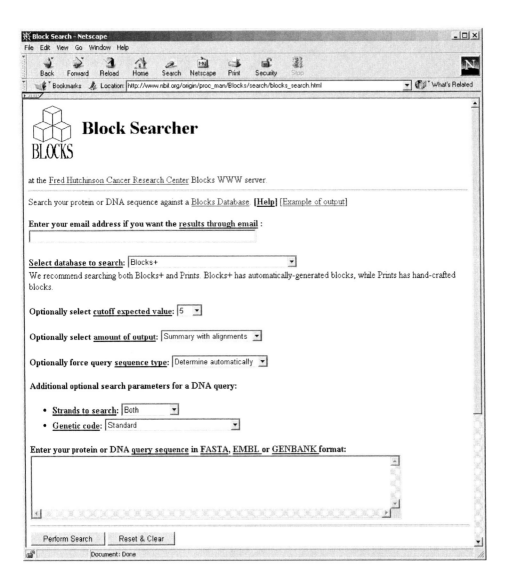

Figure 9.32. The Blocks Searcher sequence entry page.

via a graphical display. The program compares query sequences with the SMART database and the PFam multiple alignment database using the HMMER package (Eddy *et al.*, 1995) (see also Table 9.14). The SMART database consists of more than 250 signaling and extracellular domains extensively annotated with respect to cellular localization, species distribution, functional class, tertiary structure and functionally

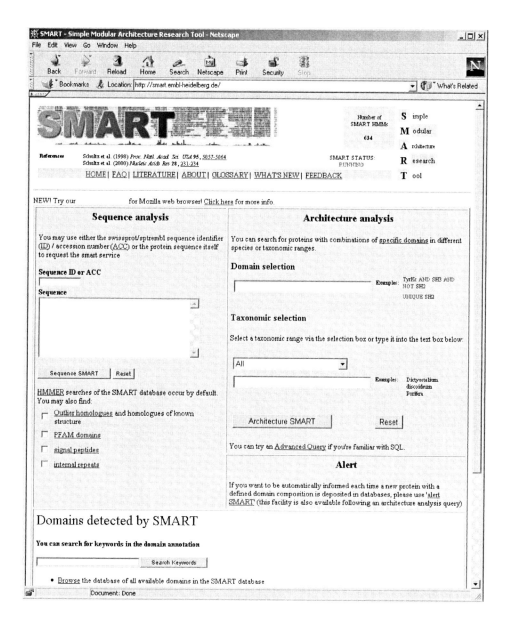

Figure 9.33. The SMART WWW home page.

important residues.

The SMART tool, while performing a sequence analysis, also allows searching for signal peptides, coiled coils, low complexity regions and transmembrane helices in the query sequence.

The SMART WWW home page is reported in Figure 9.33. Both sequence analysis and architecture analysis can be performed. For sequence analysis the SWISSPROT/TREMBL identifier or accession number of the query sequence can be typed or pasted in the 'Sequence ID or ACC' box. Otherwise the protein sequence itself can be pasted in the text window. Four options for the search are available: outlier homologues and homologues of known structure, PFam domains, signal peptides and internal repeats. If no option is selected, the program will scan the SMART database.

Given a specific domain or a combination of specific domains (a list of such domains is provided in the SMART server home page by clicking the 'specific domains' link), the architecture analysis program searches proteins sequences having such domain or combination of domains. The search can be restricted to a specific organism by choosing a taxonomic range provided by the pulldown menu.

9.6 References

Altschul, S.F., Madden, T.L., Schäffer, A.A., Zhang, J., Zhang, Z., Miller, W., and Lipman, D.J. 1997. Gapped BLAST and PSI-BLAST: a new generation of protein database search programs. Nucl. Acids Res. 25: 3389-3402.

Attwood, T.K. and Beck, M.E. 1994a. PRINTS - A protein motif fingerprint database. Protein Eng. 7: 841-848.

Attwood, T.K., Beck, M.E., Bleasby, A.J., and Parry-Smith, D.J. 1994b. PRINTS - A database of protein motif fingerprints. Nucl. Acids Res. 22: 3590-3596.

Attwood, T.K., Blythe, M., Flower, D.R., Gaulton, A., Mabey, J.E., Maudling, N., McGregor, L., Mitchell, A., Moulton, G., Paine, K., and Scordis, P. 2002. PRINTS and PRINTS-S shed light on protein ancestry. Nucl. Acids Res. 30. In press.

Attwood, T.K., Payne, A.W.R., Michie, A.D., and Parry-Smith, D.J. 1997. A colour interactive editor for multiple alignments -CINEMA. EMBnet News. 3.

Bairoch, A.1991. PROSITE: a dictionary of sites and patterns in proteins. Nucl. Acids Res. 19 Suppl.: 2241-2245.

Bairoch A. and Apweiler R. 2000. The SWISS-PROT protein sequence database and its supplement TrEMBL in 2000. Nucl. Acids Res. 28: 45-48.

Baldi, P., Chauvin, Y., Hunkapillar, T., and McClure, M. 1994. Hidden Markov models of biological primary sequence information. Proc. Natl. Acad. Sci. USA. 91: 1059-1063.

Barker, W.C., Garavelli, J.S., Hou, Z., Huang, H., Ledley, R.S., McGarvey, P.B., Mewes, H-W., Orcutt, B.C., Pfeiffer, F., Tsugita, A., Vinayaka, C.R., Xiao, C., Yeh, L-S.L., and Wu, C. 2001. Protein Information Resource: a community resource for expert annotation of protein data. Nucl. Acids Res. 29: 29-32.

Barton GJ. 1993. ALSCRIPT: a tool to format multiple sequence alignments. Protein Eng. 6: 37-40.

Corpet, F.1988. Multiple sequence alignment with hierarchical clustering. Nucl. Acids Res.16: 10881-10890.

Dayhoff, M. 1978. Atlas of Protein Sequence and Structure. Volume 5. Suppl. 3. pp 345-358. Natl. Biomed. Res. Found., Washington

Eddy, S., Mitchison, G., and Durbin, R. 1995. Maximum discrimination hidden Markov models of sequence consensus. J. Comput. Biol. 2: 9-23.

Eddy, S. 1995. Multiple alignment using hidden Markov models. In: Proc. Int. Conf. on Intelligent Systems for Molecular Biology pp. 114-120, Cambridge, England: AAAI/MIT Press.

Gribskov, M., McLachlan, A. D., and Eisenberg, D. 1987. Profile analysis: Detection of distantly related proteins. Proc. Natl. Acad. Sci. USA. 84: 4355-4358.

Grundy, W.N., Bailey, T.L., Elkan, C.P., and Baker, M.E. 1997. Meta-MEME: Motif-based Hidden Markov Models of Protein Families. Computer Appli. Biosci. 13: 397-406.

Henikoff, S., Pietrokovski, S., and Henikoff, J.G. 1998. Superior performance in protein homology detection with the Blocks Database servers. Nucl. Acids Res. 26: 309-312.

Henikoff, S. and Henikoff, J.G. 1991. Automated assembly of protein blocks for database searching. Nucl. Acids Res.19: 6565-6572.

Henikoff, S. and Henikoff, J.G. 1992. Amino acid substitution matrices from protein blocks. Proc. Natl. Acad. Sci. USA. 89: 10915-10919.

Higgins, D.G., Thompson, T.D., and Gibson, T.J. 1996. Using Clustal for multiple sequence alignments. Meth. Enzymol. 266: 383-402.

Parry-Smith, D.J., Payne, A.W.R, Michie, A.D., and Attwood, T.K. 1998. CINEMA - A novel colour interactive editor for multiple alignments. Gene. 221: GC57-63.

Pearson, W. R. and Lipman D. J. 1988. Improved tools for biological sequence analysis. Proc. Natl. Acad. Sci. USA. 85: 2444- 2448.

Pearson W. R. 1990. Rapid and Sensitive Sequence Comparison with FASTP and FASTA. Meth. Enzymol. 183: 63- 98.

Pietrokovski, S., Henikoff, J.G., and Henikoff, S. 1996. The Blocks database—a system for protein classification. Nucl. Acids Res. 24: 197-200.

Pietrokovski, S., Henikoff, J.G., and Henikoff, S. 1998. Exploring protein homology with the Blocks server. Trends Genet. 14: 162-163.

Ponting, C.P., Shultz, J., Milpetz, F., and Bork, P. 1999. SMART: identification and annotation of domains from signaling and extracellular protein sequences. Nucl. Acids Res. 27: 229-232.

Rabiner, L. R. 1989. A tutorial on hidden Markov models and selected applications in speech recognition. Proc. IEEE. 77: 257-286.

Shultz, J., Milpetz, F., Bork, P., and Ponting, C.P. 1998. SMART, a simple modular architecture research tool: Identification of signaling domains. Proc. Natl. Acad. Sci. USA. 95: 5857-5864.

Sonnhammer, E.L., Eddy, S.R., and Durbin, R. 1997. Pfam: A comprehensive database of protein domain families based on seed alignments. Protein. 28: 405-420.

Sonnhammer, E.L., Eddy, S.R.,Birney, E., Bateman, A., and Durbin, R. 1998. Pfam: Multiple sequence alignments and HMM-profiles of protein domains. Nucl. Acids Res. 26: 320-322.

10

From Sequence to Structure: an Easy Approach to Protein Structure Prediction

Fabrizio Ferré

Contents

From: *The Internet for Cell and Molecular Biologists: Current Applications and Future Potential*
ISBN 1-898486-32-8 © 2002 Horizon Scientific Press, Wymondham, UK

Abstract

The analysis of the three-dimensional structure of a protein can be very helpful in the design of experimental procedures aimed at the understanding of protein function. Experimental techniques as X-ray diffraction and Nuclear Magnetic Resonance are used to determine protein structures that are then stored in freely accessible databases. Molecular graphics software are also freely or commercially available to examine these structures. The protein structure generally depends only on the primary structure and on environmental conditions. Extrinsic factors, such as chaperones or the creation of disulfide bridges, may assist the folding process but are often not essential to it. Consequently, the protein three-dimensional structure may in principle be inferred by the sequence itself. While the experimental procedures to determine the protein three-dimensional structure are becoming faster and more reliable, the number of known sequences exceeds by far the number of known structures. Several methods have been developed to predict the protein structure from the sequence, and a number of them are freely available on the internet and easy to use. Modeling by homology is the more reliable method to predict protein structure: it is based on the assumption that, if two proteins share a high (or reasonably high) sequence identity, their 3D structure will also be similar (or reasonably similar) with good reliability.

10.1 Principles of protein structure

10.1.1 Introduction

The knowledge of protein structure helps the molecular biologist in tasks such as inferring the protein function or binding ability, designing mutagenesis and protein engineering experiments, performing rational drug design, and designing novel proteins. Moreover, the analysis and determination of protein structure appears to be of primary importance given the various genome projects undertaken, allowing in principle tasks such as mapping the function of proteins in metabolic pathways for whole genomes and deducing evolutionary relationships. The protein folding problem is therefore of primary interest in molecular biology and has attracted many scientists, but it mostly remains unsolved.

In the last few years whole-genome sequencing projects brought to the research community a huge amount of sequence data; the current increasing deficit in protein structures with respect to protein sequences has stimulated the establishment and financial support of large-scale pilot projects for protein structure determination.

10.1.1.1 Protein structure

Proteins are polymers of repeating units, amino acids, bound together by a partially double bond, the peptide bond. It is possible to identify a repetitive main-chain and a sequence of side-chains; each one of the twenty different amino acids that represent the natural repertoire in proteins has its own side-chain that differs from the other side-chains in physical and chemical properties. The sequence of the amino acids, or residues, is called the primary structure (1-D). Each protein chain folds in space to form the three-dimensional structure, which in general is uniquely determined by its amino acid sequence; biological function is strictly dependent on protein 3-D structure. The interactions among side chains, main chain, solvent and ligands determine the conformation energy. It is established that most proteins are able to assume the most energetically favourable conformation without the aid of any external factor. The three-dimensional conformation is thus uniquely determined by the amino acid sequence and corresponds to the energy minimum at the given environmental conditions. While side-chains may have polar or charged chemical groups, most of them have a hydrophobic nature. One of the main forces that determine the protein fold derives from the packing of the hydrophobic side-chains inside the molecule. Obviously, bringing these side-chains far from the polar solvent implies that the main chain, which has polar characteristics, must be buried in a hydrophobic context. To obtain an energetically favorable conformation, the main chain polar groups are involved in series of hydrogen bonds that lead to the formation of regular structures, known as secondary structure elements (SSE). Such organization allows the formation of compact structures with a hydrophobic core and a hydrophilic surface. The regular structures that can be found in proteins have a shape of helices or sheets. The alpha-helix, the most abundant helicoidal structure, is stabilized by H-bonds between the CO group of a given n residue and the NH of the $n + 4$ residue, leading to a periodic structure with 3.6 residues per helical turn. The other most abundant secondary structure element is the beta-sheet, formed by parallel or anti-parallel beta-strands that make H-bonds between the CO groups of a strand residues and the NH groups of the near strand residues. Secondary structure elements are linked together by loops presenting a non-regular structure; these loops often reach the protein surface and are solvent exposed, and functional residues are frequently localized on them.

A protein fold can be described as the rotation angles around the main chain bonds (N–Calpha angle, called psi, Calpha–C angle, called phi and the peptide bond angle, omega, whose values are

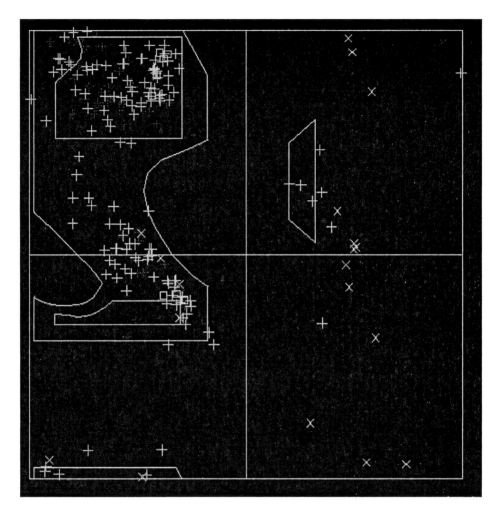

Figure 10.1. Ramachandran plot. Ramachandran plot of the alpha-catalytic subunit of the mouse cAMP-dependent protein kinase (PDB code 1jbp), calculated using the WHAT IF web server (URL http://www.cmbi.kun.nl:1100/WIWWWI).

restricted due to the partial double-bond character). The psi and phi angles value are limited only by steric collisions; some combinations of psi and phi are not allowed, depending on the residue side-chain stereo-chemical constrains. Plotting psi and phi in a Ramachandran diagram (shown in Figure 10.1) two allowed regions can be identified, one corresponding to values near psi = -57° and phi = -47° and the other to psi = -125° and phi = 125°, for a *trans* conformation of the peptide bond (omega = 180°). These two regions correspond to alpha-helices and beta-sheets, the two most common secondary structure elements. Side-chain internal rotation angles are called chi angles; different combinations of the chi angles for a given side-chain produce different side-chain conformations called rotamers, usually limited to a set of preferred structures. The rotamer conformation adopted by a side-chain is also influenced by the main-chain conformation.

The classical experiments of Anfinsen *et al.* (Anfinsen *et al.*, 1961; Anfinsen, 1973) proved that, in general, enzymes can be denatured and then refolded without significant loss of enzymatic activity. In the cell, the folding process is generally assisted by helper proteins (the chaperones), which may also correct misfolds. However, they are not essential to attain the minimum energy configuration.

10.1.1.2 Techniques for the experimental determination of protein structure

Two major techniques are widely used for protein and nucleic acid three-dimensional structure determination: X-ray diffraction and Nuclear Magnetic Resonance (NMR). Electron diffraction techniques are used for membrane protein structure determination.

X-ray diffraction

A protein purified and crystallized can diffract an X-ray beam in a pattern determined by the arrangement of the molecules in the crystal cells. Thus it is possible to obtain an electron density map of the protein from the diffraction pattern and from other information (i.e. the diffracted X-ray phases). The initial model obtained can then be refined with computational tools. The crystal quality (i.e. how much a crystal has an ordered structure) determines the model resolution, which is measured in Angstroms (the lower the value, the better the resolution). At low resolutions (greater than 5 Å) the overall protein shape can be determined, and in some cases it is possible to identify secondary structure elements (helices appear as cylindrical electronic density map regions). At average resolution (about 3 Å) it is possible to trace the main chain pattern and, if the primary structure is known (as it is often the case), it is possible to build up the side-chains over the backbone. At a 2 Å resolution side-chains are clearly identifiable, and at 1 Å each protein atom corresponds to a distinct electron density map spot.

The great majority of known structures has been determined by this technology, that has been greatly improved over the years: structures are now frequently determined with a resolution about 1.5 Å. The major limitations are due to the difficulty in obtaining a suitable crystal for many proteins of interest.

Nuclear Magnetic Resonance

The atomic nucleus of some isotopes (as ^1H, ^{13}C, ^{15}N and ^{31}P) has a property known as magnetic spin; if the molecule is immersed in a magnetic field, the spin of these isotopes aligns to the field direction, reaching an equilibrium state. Applying a radio-frequency (RF) impulse these nuclei can reach a higher energy level, and then fall back to the equilibrium emitting an RF impulse. These data are related to the nucleus' molecular environment and can be collected and analyzed. From these signals information about the molecular structure can be deduced: interactions between atom pairs at a distance shorter than 5 Å produce signal peaks (Nuclear Overhauser Effects, NOEs) from which inter-atomic distances can be calculated. NMR does not require protein crystallization, and can be applied to proteins in solution: for this reason a single NMR experiment produces a set (usually 20-30) of slightly different structures, reflecting protein dynamic behavior.

Both X-ray crystallography and NMR are technologies with known advantages and disadvantages. The requirement for a protein crystallization step for X-ray crystallography can be very restrictive, and the protein conformation in the crystal can be an artificial one, never adopted in a cellular environment. It has however been observed that crystallized enzymes usually maintain their catalytic activity. Nevertheless, the structure can be artificially altered in some small regions. The major limitations affecting the NMR technique are related to the protein size: up to now the upper size limit seems to be about 300 residues.

Table 10.1. Protein Data Bank contents. Structures contained in the PDB, updated at 13th November 2001 release. The great majority of structures have been determined by X-ray diffraction, while the number of NMR experiments is rapidly growing; a few models obtained by theoretical calculation are also stored.

		Molecule type				
		Proteins, peptides and viruses	Protein/nucleic acids complexes	Nucleic acids	Carbohydrates	Total
Experimental technique	X-ray diffraction	12488	601	591	14	13654
	NMR	2056	78	407	4	2545
	Theoretical models	301	23	23	0	347
	Total structures	14805	702	1021	18	16546

10.1.2 Structures databases

Protein structures determined so far are stored in several databases, freely accessible on the web; some of them are annotated collections of structures, while others try to build a protein classification based on structural and functional features. Here is an overview of the most popular databases available through the World Wide Web.

10.1.2.1 The Protein Data Bank and PDBSum

The Protein Data Bank, known as PDB (main URL http://www.rcsb.org/pdb/, several mirror web sites exist), is a protein structure database managed by the Research Collaboration for Structural Bioinformatics (RCSB), a consortium composed of members of Rutgers University (New Jersey, USA), the National Institute of Standards and Technology (NIST) and the San Diego Supercomputer Center (SDSC).

The database holds protein structures but also nucleic acid and carbohydrate structures, determined by X-ray diffraction, NMR or

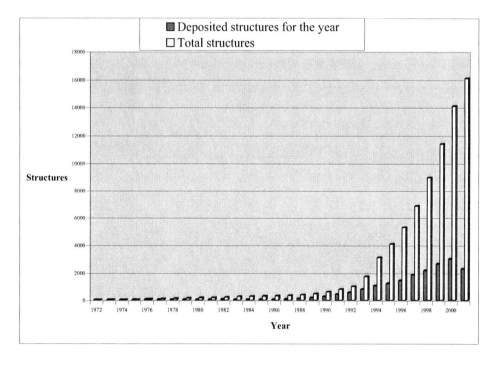

Figure 10.2. Growth of the PDB database. A dramatic growth in the number of structures deposited per year is clearly detectable in the last ten years, due to the improvements in experimental techniques.

other less used experimental techniques; a few theoretical models are also stored. It was established in 1971 at the Brookhaven National Laboratories, holding seven structures; the latest update (at 13th November 2001) includes 16546 structures (Table 10.1), and data are updated once a week. The number of deposited structures per year has increased dramatically from the early nineties (Figure 10.2), given the improvement in the experimental techniques. Nevertheless, there is a certain degree of redundancy, as some protein structures have been re-determined in different conditions or at higher resolution, and in some cases the structure of very closely related proteins has been determined.

Each PDB entry contains the atomic three-dimensional coordinates and other information, regarding the protein itself (name, organism, sequence and bibliographic references) and experimental details and quality. There are three different query interfaces that can

```
HEADER    TRANSFERASE                           06-JUN-01   1JBP        SEQRES  12 E  350  ARG PHE TYR ALA ALA GLN ILE VAL LEU THR PHE GLU TYR
TITLE     CRYSTAL STRUCTURE OF THE CATALYTIC SUBUNIT OF CAMP-           SEQRES  13 E  350  LEU HIS SER LEU ASP LEU ILE TYR ARG ASP LEU LYS PRO
TITLE    2 DEPENDENT PROTEIN KINASE COMPLEXED WITH A SUBSTRATE          SEQRES  14 E  350  GLU ASN LEU LEU ILE ASP GLN GLN GLY TYR ILE GLN VAL
TITLE    3 PEPTIDE, ADP AND DETERGENT                                   SEQRES  15 E  350  THR ASP PHE GLY PHE ALA LYS ARG VAL LYS GLY ARG THR
COMPND    MOL_ID: 1;                                                    SEQRES  16 E  350  TRP TPO LEU CYS GLY THR PRO GLU TYR LEU ALA PRO GLU
COMPND   2 MOLECULE: CAMP-DEPENDENT PROTEIN KINASE, ALPHA-CATALYTIC     SEQRES  17 E  350  ILE ILE LEU SER LYS GLY TYR ASN LYS ALA VAL ASP TRP
COMPND   3 SUBUNIT;                                                     SEQRES  18 E  350  TRP ALA LEU GLY VAL LEU ILE TYR GLU MET ALA ALA GLY
COMPND   4 CHAIN: E;                                                    SEQRES  19 E  350  TYR PRO PRO PHE PHE ALA ASP GLU PRO ILE GLN ILE TYR
COMPND   5 SYNONYM: PKA C-ALPHA;                                        SEQRES  20 E  350  GLU LYS ILE VAL SER GLY LYS VAL ARG PHE PRO SER HIS
COMPND   6 EC: 2.7.1.37;                                                SEQRES  21 E  350  PHE SER SER ASP LEU LYS ASP LEU LEU ARG ASN LEU LEU
COMPND   7 ENGINEERED: YES;                                             SEQRES  22 E  350  GLN VAL ASP LEU THR LYS ARG PHE GLY ASN LEU LYS ASN
COMPND   8 MOL_ID: 2;                                                   SEQRES  23 E  350  GLY VAL ASN ASP ILE LYS ASN HIS LYS TRP PHE ALA THR
COMPND   9 MOLECULE: CAMP-DEPENDENT PROTEIN KINASE INHIBITOR,           SEQRES  24 E  350  THR ASP TRP ILE ALA ILE TYR GLN ARG LYS VAL GLU ALA
COMPND  10 MUSCLE/BRAIN FORM;                                           SEQRES  25 E  350  PRO PHE ILE PRO LYS PHE LYS GLY PRO GLY ASP THR SER
COMPND  11 CHAIN: S;                                                    SEQRES  26 E  350  ASN PHE ASP ASP TYR GLU GLU GLU GLU ILE ARG VAL SEP
COMPND  12 FRAGMENT: RESIDUES 5-24;                                     SEQRES  27 E  350  ILE ASN GLU LYS CYS GLY LYS GLU PHE THR GLU PHE
COMPND  13 SYNONYM: PKI-ALPHA;                                          SEQRES   1 S   20  THR THR TYR ALA ASP PHE ILE ALA SER GLY ARG THR GLY
COMPND  14 ENGINEERED: YES;                                             SEQRES   2 S   20  ARG ARG ALA SER ILE HIS ASP
COMPND  15 MUTATION: YES                                                MODRES 1JBP SEP E   10  SER  PHOSPHOSERINE
SOURCE    MOL_ID: 1;                                                    MODRES 1JBP TPO E  197  THR  PHOSPHOTHREONINE
SOURCE   2 ORGANISM_SCIENTIFIC: MUS MUSCULUS;                           HET     SEP E  10      10
SOURCE   3 ORGANISM_COMMON: MOUSE;                                      HET     TPO E 197      11
SOURCE   4 EXPRESSION_SYSTEM: ESCHERICHIA COLI;                         HET     SEP E 338      10
SOURCE   5 EXPRESSION_SYSTEM_COMMON: BACTERIA;                          HET     ADP   381      27
SOURCE   6 MOL_ID: 2;                                                   HET     OCT   382       8
SOURCE   7 SYNTHETIC: YES;                                              HETNAM     SEP PHOSPHOSERINE
SOURCE   8 OTHER_DETAILS: THE PEPTIDE WAS CHEMICALLY SYNTHESIZED. THE   HETNAM     TPO PHOSPHOTHREONINE
SOURCE   9 SEQUENCE OF THE PEPTIDE IS NATURALLY FOUND IN MUS MUSCULUS   HETNAM     ADP ADENOSINE-5'-DIPHOSPHATE
KEYWDS    PROTEIN-SUBSTRATE COMPLEX                                     HETNAM     OCT N-OCTANE
EXPDTA    X-RAY DIFFRACTION                                             HETSYN     SEP PHOSPHONOSERINE
AUTHOR    MADHUSUDAN,E.A.TRAFNY,N.H.XUONG,J.A.ADAMS,L.F.TEN EYCK,       HETSYN     TPO PHOSPHONOTHREONINE
AUTHOR   2 S.S.TAYLOR,J.M.SOWADSKI                                      FORMUL   1  SEP    2(C3 H8 N1 O6 P1)
REVDAT   1  27-JUN-01 1JBP    0                                         FORMUL   1  TPO    C4 H10 N1 O6 P1
JRNL        AUTH   MADHUSUDAN,E.A.TRAFNY,N.H.XUONG,J.A.ADAMS,           FORMUL   3  ADP    C10 H15 N5 O10 P2
JRNL        AUTH 2 L.F.TEN EYCK,S.S.TAYLOR,J.M.SOWADSKI                 FORMUL   4  OCT    C8 H18
JRNL        TITL   CAMP-DEPENDENT PROTEIN KINASE: CRYSTALLOGRAFIC       FORMUL   5  HOH    *168(H2 O1)
JRNL        TITL 2 INSIGHTS INTO SUBSTRATE RECOGNITION AND              ATOM      1  N   GLY E   9      -4.342 -22.211 -16.053  1.00 99.99           N
JRNL        TITL 3 PHOSPHOTRANSFER                                      ATOM      2  CA  GLY E   9      -5.385 -22.161 -15.037  1.00 53.23           C
JRNL        REF    PROTEIN SCI.              V.   3   176 1994          ATOM      3  C   GLY E   9      -5.805 -19.810 -15.496  1.00 99.99           C
JRNL        REFN   ASTM PRCIEI  US ISSN 0961-8368                       ATOM      4  O   GLY E   9      -6.240 -20.877 -15.068  1.00 28.58           O
                                                                       HETATM    5  N   SEP E  10      -7.472 -21.030 -14.580  1.00 41.15           N
DBREF  1JBP E    1   350  SWS    P05132   KAPA_MOUSE       1    350     HETATM    6  CA  SEP E  10      -8.480 -20.008 -14.476  1.00 25.96           C
DBREF  1JBP S  361   380  SWS    P27776   IPKA_MOUSE       5     24     HETATM    7  CB  SEP E  10      -9.666 -20.624 -13.771  1.00 36.66           C
SEQADV 1JBP SEP E   10  SWS  P05132      SER     10 MODIFIED RESIDUE    HETATM    8  OG  SEP E  10     -10.934 -20.172 -14.244  1.00 99.99           O
SEQADV 1JBP GLU E  242  SWS  P05132      GLN    242 CONFLICT            HETATM    9  C   SEP E  10      -7.990 -18.821 -13.665  1.00 29.23           C
SEQADV 1JBP SEP E  338  SWS  P05132      SER    338 MODIFIED RESIDUE    HETATM   10  O   SEP E  10      -8.196 -17.684 -14.080  1.00 99.99           O
SEQADV 1JBP ALA S  376  SWS  P27776      ASN     20 ENGINEERED          HETATM   11  P   SEP E  10     -12.244 -20.797 -13.532  1.00 99.99           P
SEQRES   1 E  350  GLY ASN ALA ALA ALA ALA LYS LYS GLY SEP GLU GLN GLU HETATM   12  O1P SEP E  10     -11.873 -21.100 -12.117  1.00 67.87           O
SEQRES   2 E  350  SER VAL LYS GLU PHE LEU ALA LYS ALA LYS GLU ASP PHE HETATM   13  O2P SEP E  10     -12.543 -22.079 -14.225  1.00 79.76           O
SEQRES   3 E  350  LEU LYS LYS TRP GLU THR PRO SER GLN ASN THR ALA GLN HETATM   14  O3P SEP E  10     -13.316 -19.774 -13.635  1.00 99.99           O
SEQRES   4 E  350  LEU ASP GLN PHE ASP ARG ILE LYS THR LEU GLY THR GLY ATOM     15  N   GLU E  11      -7.393 -19.108 -12.488  1.00 40.91           N
SEQRES   5 E  350  SER PHE GLY ARG VAL MET LEU VAL LYS HIS LYS GLU SER ATOM     16  CA  GLU E  11      -6.884 -18.069 -11.592  1.00 99.99           C
SEQRES   6 E  350  GLY ASN HIS TYR ALA MET LYS ILE LEU ASP LYS GLN LYS ATOM     17  C   GLU E  11      -5.858 -17.180 -12.266  1.00 99.99           C
SEQRES   7 E  350  VAL VAL LYS LEU LYS GLN ILE GLU HIS THR LEU ASN GLU ATOM     18  O   GLU E  11      -5.882 -15.947 -12.144  1.00 36.61           O
SEQRES   8 E  350  LYS ARG ILE LEU GLN ALA VAL ASN PHE PRO PHE LEU VAL ATOM     19  CB  GLU E  11      -6.425 -18.628 -10.268  1.00 47.67           C
SEQRES   9 E  350  LYS LEU GLU PHE SER PHE LYS ASP ASN SER ASN LEU TYR ATOM     20  N   GLN E  12      -4.942 -17.810 -12.988  1.00 28.19           N
SEQRES  10 E  350  MET VAL MET GLU TYR VAL ALA GLY GLY GLU MET PHE SER ATOM     21  CA  GLN E  12      -3.995 -17.033 -13.736  1.00 73.50           C
SEQRES  11 E  350  HIS LEU ARG ARG ILE GLY ARG PHE SER GLU PRO HIS ALA
```

Figure 10.3. Parts of a PDB file. The header and the coordinates for the first N-terminal residues of the PDB file 1jbp (the structure of the alpha-catalytic subunit of the mouse cAMP-dependent protein kinase) are shown. The first part (HEADER, TITLE, COMPND, SOURCE, KEYWDS, EXPDATA, AUTHOR, REVDAT, JRNL fields) reports general information about the protein analyzed, the experimental technique used and the authors of the work and the publication. Then information about the protein sequence is shown in the SEQRES part, while the HET and HETATM lines refers to ligands or post-translational modifications. Each ATOM line reports three-dimensional coordinates of a protein atom, specifying the residue to which the atom belongs.

be used to retrieve the structure of interest: the most simple is the SearchLite, that uses one or more keywords to search trough the database; the SearchField interface allows a more customizable search and offers the possibility to perform a FASTA sequence search (see Chapter 9); queries for unreleased entries can be performed using the Status interface. Once the desired structures are identified, the user can view the corresponding files in plain text or HTML format with a browser, or can download the files in different compressed or uncompressed formats (Figure 10.3). Structures can be viewed with specific software (browser plug-ins, as Chime or VRML), or through the QuickPDB java applet (see the Links Paragraph for URLs and other information); once downloaded, structure files (in PDB or mmCIF format) can be viewed with molecular graphics tools (see Paragraph 10.1.3).

The PDBsum database (URL http://www.biochem.ucl.ac.uk/bsm/pdbsum/) contains summary information and derived data on each entry in the PDB; for each chain a schematic diagram is given, showing the protein amino acid sequence, its secondary structure elements, domain composition and motifs. The derived data include PROMOTIF analyses, checking for structural motifs in the protein (Hutchinson and Thornton, 1996), PROCHECK statistics (PROCHECK is a program that checks the stereochemical quality of a structure; Laskowski *et al.*, 1993), a LIGPLOT (Wallace *et al.*, 1995) schematic representation of protein-ligand interactions, and links to several databases.

10.1.2.2 SCOP

The Structural Classification of Proteins database (SCOP, URL http://scop.mrc-lmb.cam.ac.uk/scop/) aim is to provide a description of the structural and evolutionary relationships between all proteins with known structure, by means of a classification that reflects structural and evolutionary relationships (Lo Conte *et al.*, 2000); this classification is made both by manual inspection and automated methods. The classification is hierarchical, with four major levels: class, fold, superfamily and family. The class level is referred to the protein secondary structure composition (all alpha, all beta, alpha-beta, and so on) or to other general structural features (i.e. membrane proteins, small proteins, coiled coils). Proteins that share a general structural similarity, with similar secondary element arrangements, are classified in the same 'fold' level. Proteins of the same fold category may not have a common evolutionary origin. Proteins that share a low sequence identity, but whose structure and function is similar, are placed in the same 'superfamily' level; these proteins share a plausible common evolutionary origin. At the family level the proteins clustered together are clearly evolutionarily related, with sequences at least 30% identical (nevertheless, there are cases in which a clear common ancestor is shared without high levels of sequence identity). Given the very special nature of the database (it is manually managed by an expert team), the exact position of boundaries between these levels is subjective and not exactly defined. The database can be accessed at the top of the hierarchy (the 'class' level), or through a keyword search.

10.1.2.3 CATH

The CATH database (URL http://www.biochem.ucl.ac.uk/bsm/cath_new/index.html; Orengo *et al.*, 1997) is a hierarchical classification of protein domain structures, which clusters proteins at four major levels: class (C), architecture (A), topology (T) and homologous superfamilies (H). Class describes structure content, while Architecture refers to the organization of secondary structure elements, independent of connectivities between the secondary structures. Topology refers to topological connections and numbers of secondary structures, and is equivalent to the SCOP fold level. The homologous superfamilies level contains proteins with highly similar structures and functions. The classification is made with a mixture of manual and automatic steps.

10.1.2.4 DSSP

The DSSP database (Dictionary of Protein Secondary Structure; URL http://www.cmbi.kun.nl/swift/dssp/; Kabsch and Sander, 1983) contains information about secondary structures for each PDB entry derived with the DSSP program (Kabsch and Sander, 1983). The secondary structure is reported (H for alpha-helix residues; E for beta conformation; T for residues belonging to turn and loops) and hydrogen bonds are indicated for each protein residue. Moreover, solvent accessibility is calculated for each residue.

10.1.2.5 DALI, FSSP and HSSP

FSSP (Fold classification based on Structure-Structure alignment of Proteins, URL http://www2.embl-ebi.ac.uk/dali/fssp/; Holm and Sander, 1996) is a database of structural alignments, which are automatically performed using the DALI program (Holm and Sander, 1997). Sequence redundancy is avoided by selecting one member for each family of proteins with at least 25% sequence homology. Selected protein structures are compared, and then clustered in groups sharing a similar fold. The DALI program, given two sets of coordinates, is able to find the biggest common substructures and then to superpose the corresponding residues.

The DALI web server (URL http://www2.embl-ebi.ac.uk/dali/) is a network service that allows the user to submit a structure to be compared against a representative subset of structures from the Protein Data Bank. Moreover, the DALI server contains a structural domain classification (the DALI Domain Dictionary), determined automatically with the criteria of recurrence and compactness (Holm and Sander, 1998).

The Homology-derived Structures of Proteins database, known as the HSSP database (URL http://www.cmbi.kun.nl/swift/hssp/; Sander and Schneider, 1991), is a database of multiple alignments obtained comparing all the available SWISS-PROT sequences (see Chapter 9) to each protein of known structure. Sequences sharing at least 30% sequence identity with the protein of known structure are reported. For each known structure the database reports the aligned sequences, the secondary structure, the solvent accessibility and the sequence profile (see Chapter 9).

10.1.3 Visualization of molecular structures: molecular graphics tools

PDB or mmCIF format structure files can be downloaded and used to visualize protein structures. Molecular graphic programs allow the user to view, rotate, translate and manipulate in different ways the protein structure. They can also be used to produce nice pictures for papers and scientific presentations.

The basic protein structure representation describes each atom as a single point in the space and each bond as a line (*wire-frame* representation). This representation is useful if one is interested in localizing the position of specific residues in the protein structure, but the recognition of more general features can be difficult. Another type of representation is the protein 'trace' or 'backbone', that can be obtained tracing a line or a ribbon that follows the Calpha atoms of each residue; this allows the easy identification of secondary structure elements and gives a clear view of the protein topology. In the so-called CPK (space-filling) representation, each atom is described as a sphere; the radius of each sphere is the Van der Waals atomic radius (VdW representation).

For each structure the molecular surface can be calculated as the solvent-accessible portion of the protein; different methods of surface determination can be used, e.g. Connolly's method (Connolly, 1983), GRASP (Nicholls *et al.*, 1991) and Naccess (made by S. Hubbard and J. Thornton), that are very useful in the analysis of the protein interaction surfaces with ligands or other proteins.

While there are complete and sophisticated commercial molecular graphics tools (e.g. Insight II, Dayringer *et al.*, 1986), many are freely available and can be downloaded from the web and installed locally.

10.1.3.1 RasMol

Probably one of the most famous and easy to use is RasMol (URL http://www.bernstein-plus-sons.com/software/rasmol/; Sayle and Milner-White, 1995) that is able to open structure files in different formats and allows the representation of proteins, nucleic acids and small organic compounds. The latest version (2.10.1) is freely available for Windows, MacOS, Unix and Linux operating systems. The program has been developed at the University of Edinburgh's Biocomputing Research Unit and the Biomolecular Structures Group at Glaxo Research and Development, Greenford, UK. The loaded molecule can be shown as wire-frame, cylinder stick bonds, alpha-carbon trace, space-filling spheres, macromolecular ribbons, hydrogen bonding and dot surface: in Figure 10.4 different possible representations of the cAMP-dependent protein kinase (PDB code 1jbp) are shown. Atoms may also be labeled with arbitrary text strings. Different parts of the molecule may be displayed and coloured independently on the rest of the molecule. A Command Line window allows the user to insert instructions by typing them; moreover, RasMol can read a prepared list of commands from a script file (this feature can be particularly useful to regenerate the current image). The rendered image may be saved in a variety of formats (raster and vector PostScript, GIF, PPM, BMP, PICT, Sun rasterfile). Some software tools, based on RasMol, can be used to visualize structures through a browser, as Chime (a Netscape plug-in by MDLI Inc.) or the Java applet WebMol (Walther, 1997). These software modules can be useful to give a quick look at a structure. Detailed instructions and tutorials for the use of RasMol

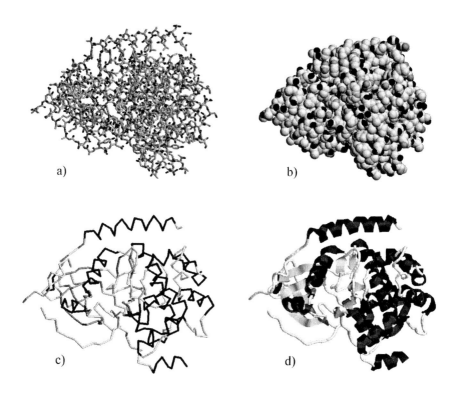

Figure 10.4. Molecule visualization with the RASMOL program. Four different protein representations of the alpha-catalytic subunit of the mouse cAMP-dependent protein kinase (PDB code 1jbp) obtained with the RASMOL software: a) wireframe; b) space-filling; c) trace; d) ribbon.

can be found on the web (e.g. at http://info.bio.cmu.edu/Courses/ BiochemMols/RasMolTutorial/RasTut.html or http://www.umass.edu/ microbio/rasmol/rastut.htm).

10.1.3.2 SwissPDB Viewer

The SwissPDB Viewer (URL http://www.expasy.ch/spdbv/; Guex and Peitsch, 1997), though easy to use, allows complex tasks such as homology modelling, being tightly linked to Swiss-Model, an automated homology modelling server (see paragraph 10.2.2.2). SwissPDB Viewer is freely available for Windows, MacOS, Unix and Linux OS. This software allows the analysis of several molecules at the same time: different proteins can be superimposed in order to deduce structural alignments and compare functional regions. It is possible to compute amino acid mutations, H-bonds, angles and distances between atoms. To simulate mutations, a rotamer library, reporting the most probable side chain torsion angles depending on the backbone conformation, is available in order to change amino acid sidechains. Structural parameters, such as the torsion angles of amino acids and heteroatoms, can be altered as well as the backbone omega, phi and psi angles. A version of the GROMOS program (van Gunsteren *et al.*, 1994) is included, which is a force field that can be used for the evaluation of the energy of a structure, allowing the repair of distorted geometries through energy minimization (this can be useful when a mutation has been introduced or when torsion has been applied to some angles, or, more generally, whenever the structure has been manually distorted). Different protein representations obtained with SwissPDB Viewer are shown in Figure 10.5.

10.1.3.3 VRML

VRML stands for Virtual Reality Modeling Language, the open standard for virtual reality on the Internet. VRML files can represent 3D computer-generated graphics, 3D sound, and hypermedia links. VRML is useful for a variety of applications, including scientific visualization. A VRML file is associated with each PDB file: with an

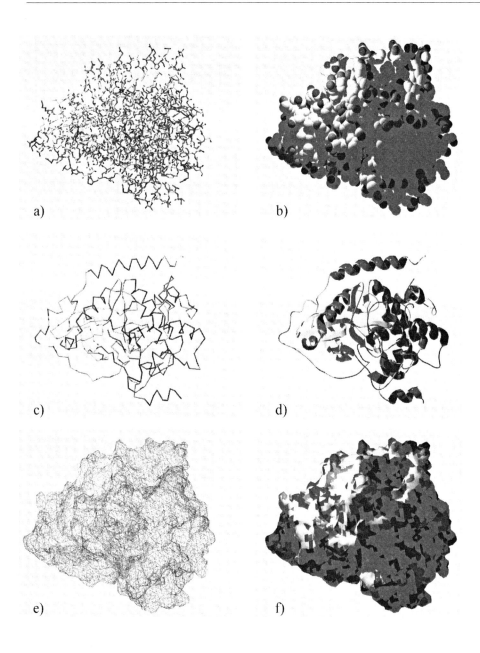

Figure 10.5. Molecule visualization with the SwissPDBViewer program. Six different protein representations of the alpha-catalytic subunit of the mouse cAMP-dependent protein kinase (PDB code 1jbp) obtained with the SwissPDBViewer software: a) wireframe; b) space-filling; c) trace; d) ribbon; e) transparent molecular surface; f) solid molecular surface. The pictures have been rendered with the POV-Ray (Persistence Of Vision Raytracer) software, freely downloadable at http://www.povray.org/.

appropriate browser's VRML v2.0 plug-in (such as the Cosmo Player, freely available at http://www.cai.com/cosmo/ for Windows and MacOS operating systems) it is possible to view, rotate and translate the structure directly through the browser.

10.1.4 Protein structure comparison

The conservation of protein sequences may reflect an evolutionary or a functional relationship (see Chapter 9); in other words, proteins sharing a certain degree of sequence similarity are supposed to share a common ancestor or to have converged to a similar arrangement of residues. These assumptions can be made also for protein structure similarities. Mutations sometimes reflect conformational changes: surface residues not directly involved in protein function are usually free to mutate, while mutations of buried residues more frequently produce structural distortions; moreover, different sequences may fold in a similar three-dimensional structure. Proteins with divergent sequence may share a similar structure, while proteins with similar sequence may adopt a different fold; then, tools developed for sequence comparison may fail in detecting such relationships.

As sequence comparison may be a very useful tool in the identification of protein function, structure comparison may be useful when sequence divergence did not involve structural distortions. Moreover, structure comparison is a fundamental tool in the classifications of proteins (see the SCOP, CATH and FSSP databases described in paragraph 10.1.2) and in the evaluation of protein structure prediction tools reliability (paragraph 10.2.5). Non-trivial similarities have been detected by manual inspection in the early years, but with the continuous growth of the number of available structures, effective tools for structure comparison became very useful resources.

The most used parameter to evaluate the similarity of two structures is the root mean square deviation, known as rms or rmsd, defined as the average of the distance vectors between two sets of coordinates, given an optimal superposition. Other parameters have been proposed, such as the protein structural distance, but the rmsd is still the most used.

In principle, sequence and structure comparison involves the same approach: the proteins to be compared first must be aligned. To align two structures, one of them should be rotated and translated as a rigid body and a three-dimensional superposition must be obtained. Since a superposition can be performed only between objects composed of the same number of atoms, when comparing proteins with non-identical sequences, only Calpha atoms can be used. Protein loops, where insertions and deletions are usually located, cannot be considered. This rigid body approach may be effective when comparing similar structures, but can easily fail in detecting local substructure similarities. Different approaches were developed to overcome these limitations: dividing the structures into fragments (Vriend and Sander, 1991), using distance matrices (Holm and Sander, 1993) or applying dynamic programming (Taylor, 1999). In the first case, a widely used trick consists of dividing the structures into secondary structure elements, which can used as equivalent regions to start the alignment. Distance matrices can be defined as the intramolecular geometric relationships among objects such as secondary structure elements or Calpha atoms, with the advantage of being independent of the three-dimensional orientation of the coordinates. Dynamic programming algorithms are tools for finding the global optimum for a given problem, and have been successfully used in sequence alignment procedures but also in structure alignment.

Different on-line resources are now available, and the most used (DALI, SARF2, TOP) are described in paragraph 10.4.1; usually the procedure requires the upload of the structures to be aligned from the user computer or alternatively the PDB codes can be specified.

10.2 Protein Structure Prediction

As shown in the previous paragraphs, a lot of useful information can be derived from analysis of protein structure. While protein structure determination techniques are becoming faster and more reliable, the number of known sequences still exceeds by far the number of known structures. Different approaches can be used to model the unknown structure of a protein. In this chapter a general strategy will be described (and summarized in Figure 10.6), while a more detailed sketch of each different method, and the related resources available on-line, can

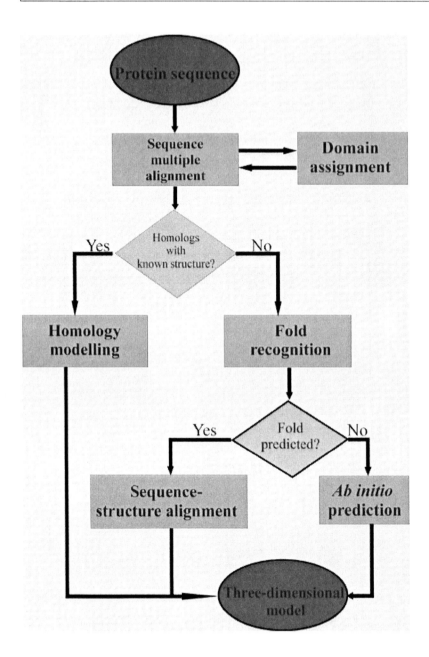

Figure 10.6. Protein structure prediction strategy. A general strategy to build a protein model structure starting from the sequence is shown. If the target sequence has one or more homologs in the PDB, homology modeling can be attempted. Otherwise, fold recognition techniques can infer the fold adopted by the sequence and, in principle, build a sequence-structure alignment. If fold recognition fails, *ab initio* prediction techniques try to predict the structure adopted basing on physico-chemical and/or semi-empirical rules.

be found in the following sections.

The first step in a modelling procedure (going from a protein sequence to a model of the protein structure) consists of a thorough analysis of the target protein sequence. It is usually preferable to split, if possible, the target sequence into discrete functional domains and consider each one separately, using such tools as multiple alignments and/or functional domain databases (e.g. SMART; see chapter 9). Moreover, sequence analysis can yield additional information that can be useful in the choice of the modelling strategy, such as the presence of transmembrane or coiled-coils regions.

More importantly, a sequence comparison with the database of the sequences of the proteins of known structure (PDB) may highlight the presence of putative good templates for a procedure known as "homology" or "comparative" modelling (see Paragraph 10.2.2). Modelling by homology is up to now the most reliable method of protein structure prediction: it is based on the assumption that if two proteins share a high (or reasonably high) sequence identity, their 3D structure is also similar (or reasonably similar) with good reliability. It is established that a quite reliable and accurate model can be obtained when the sequence identity between the target and the most similar of the PDB sequences is higher than 70%; around 50% identity the model is still reliable, while it is not advised to try to build the model when identity falls under the 30% threshold. Different on-line tools easily allow the building of a 3D model starting only from the protein sequence without any knowledge of the procedure itself. The most used and powerful web resources are reported and discussed in paragraph 10.2.2.2.

If no suitable template protein is found, it is still possible to obtain structural information about the target sequence. Several 'fold recognition' methods are able to predict the fold adopted by the protein (Paragraph 10.2.3). Different approaches have been explored, with the common assumption that the number of folds and topologies that proteins may adopt is limited, and that unrelated proteins may share a similar fold. The 'threading' procedures use empirical or semi-empirical scoring functions to evaluate the fit of a target sequence with each member of a fold library. Threading methods are thought to be the most effective, although promising results have been recently obtained with approaches based on secondary structure prediction combined with multiple sequence alignments. When a suitable fold has been chosen, it is possible to align the sequence and the selected fold obtaining a three-dimensional model. However, it should be considered that the alignment of a sequence on a structure suggested by fold recognition methods might be inaccurate.

If fold recognition techniques fail, the last resource relies on the so-called *ab initio* methods (Paragraph 10.2.4), which consist of the prediction of protein tertiary structure in the absence of any similarity to a known structure, by simulating *in silico* the natural process of protein folding. These complex and sophisticated techniques are very expensive from the computational point of view and cannot in general be considered as accessible to the molecular and cell biologist, although a few on-line resources are now available (Paragraph 10.2.4.2).

10.2.1 Secondary structure prediction

10.2.1.1 Introduction

The prediction of protein secondary structure can be very useful in tasks such as protein engineering or mutant design, NMR protein structure determination, X-ray crystal structure validation, protein structure comparison and protein folding prediction with threading or *ab initio* techniques. Up to now several different techniques have been developed, with a reasonably high degree of reliability: starting only from a single protein sequence, the best prediction method can predict its secondary structure with an accuracy of around 65% on a residue basis; if multiple alignment information is available, the accuracy increases to 70%, while if the structural domain is known, the value can reach about 80%.

Secondary structure prediction methods can be classified into statistical, knowledge-based/physico-chemical and hybrid systems. Statistical methods are based on empirical rules, derived from the analysis of proteins with known structure, which describes the relationships between primary and secondary structure. Knowledge-based methods also are based on analysis of known structures, but the goal is to derive physical and chemical rules for secondary structures formation. Features of both statistical and physico-chemical approaches are considered in hybrid methods.

One of the most successful prediction methods is the Chou-Fasman method (Chou and Fasman, 1974a; Chou and Fasman, 1974b), based on the calculation of the statistical propensity of each residue to adopt a helical or a beta strand secondary structure conformation. The GOR method (Garnier *et al.*, 1978; Gibrat *et al.*, 1987) is based on statistical rules, but hides a more sophisticated theory, that includes relationships between different residues in different states. Some statistic methodology takes advantage of neural networks to make the system learn from known structures (Qian and Sejnowski, 1988; Rost and Sander, 1993; McGregor *et al.*, 1989). The Lim method (Lim, 1974) is based on theoretical considerations regarding globular protein packing, defining rules that must be satisfied in order to form secondary structure elements.

10.2.1.2 On the web

On-line resources

PSIPRED

PSIPRED (URL http://insulin.brunel.ac.uk/psipred/) is a protein structure prediction server that allows the user to obtain different kinds of predictions about the target sequence. The PSIPRED method itself (Jones, 1999) predicts secondary structure by means of a neural network which analyzes a position specific scoring matrix generated by PSI-BLAST; its accuracy (around 75%) is one of the highest among available methods. A java application gives a graphical representation of the results. In addition to secondary structure prediction, the user can also find a transmembrane topology prediction method (MEMSAT 2, see paragraph 10.3) and a fold recognition procedure (GenTHREADER, see paragraph 10.2.3).

SSPro

SSpro (URL http://promoter.ics.uci.edu/BRNN-PRED/) is a server for protein secondary structure prediction based on an ensemble of 11 bi-directional recurrent neural networks; the system takes as input a profile obtained from a multiple alignment, produced by an algorithm based on the PSI-BLAST procedure. The final predictions are obtained by averaging the network outputs for each residue. The method has a very high accuracy (around 78%), and is thought to be one of the most effective.

PHDsec

PHDsec (Rost, 1996) is part of the PredictProtein server (URL http://cubic.bioc.columbia.edu/predictprotein/; see the Links paragraph); a system of neural networks is used, which takes advantage from evolutionary information derived from multiple sequence alignments, reaching an average accuracy around 72%.

JPred

The JPred server (URL http://jura.ebi.ac.uk:8888/) submits the query to various different prediction methods, reporting results from each method and a consensus prediction. Methods used are DSC (King and Sternberg, 1996), PHD (Rost and Sander, 1993), NNSSP (Salamov and Solovyev, 1995), PREDATOR (Frishman and Argos, 1997), ZPRED (Zvelebil et al., 1987), MULPRED (Barton, unpublished), COILS (Lupas, 1996), MultiCoil (Wolf et al., 1997) and Jnet (Cuff and Barton, 2000); the majority of these methods have their own web server, but the combination of the predictions results in an accuracy improvement (Cuff et al., 1998) compared with the accuracy of each method.

10.2.2 Homology Modelling

10.2.2.1 Introduction

The easiest way to build protein three-dimensional models from the sequence is when the structure of one or more homologues are known, and a 3D model can be built based on the structural information derived from the homologues. This is called "homology" or "comparative" modelling. It is established that the higher the homology, the more reliable will be the model: when two proteins do share a significant sequence identity, their structure is similar. Structural differences between superimposed homologous proteins (i.e. measured as the rmsd, see paragraph 10.1.4) decrease as their sequence similarity increases (Chung and Subbiah, 1996; Chothia and Lesk, 1986; Flores *et al.*, 1993). As shown by Hilbert and co-workers (1993), when sequence identity between two proteins increases, the size of the common core region concurrently increases, as superposed proteins with sequence identity greater than 50% have at least 90% of the residues structurally conserved.

It is important to notice that up to 0.5 Å rmsd of alpha carbons occurs in independent X-ray crystallographic determinations of the same protein, while proteins with 50% sequence identity have on average 1 Å rmsd (Chothia and Lesk, 1986). Although there are exceptions to these general associations between sequence homology and structural differences, the above observations provide some tools for the assessment of accuracy at a given level of sequence similarity. Thus, the modelling procedure reliability is strictly dependent on the accuracy of the sequence alignment of the target protein to the template; trying to obtain a model when the sequence identity is lower than 30% can lead to a fake structure.

Several different methods have been developed for homology modelling, but all of them follow six similar steps: alignment of the target sequence on the template structure backbone; building of a framework structure; addition and optimization of side chains; loop building; refinement of the model; validation of the model (Moulth, 1996). These steps will now be described individually, with the aim of providing a basic overview of the modelling process. In the next section, automated methods available on-line will be described.

i) Alignment

This step plays a critical role for the overall reliability of the resulting model. For very closely related proteins (70% sequence identity or more) this is a trivial task, but when the homology is poor, the more frequently used algorithms can fail to produce an optimal alignment, and often, manual intervention can improve the alignment reliability. Obviously this can be a problem when using the automated homology modelling procedures available on the web.

ii) Building of a framework structure

If there is more than one structure with good sequence similarity to the target protein, all of them can be aligned to look for the more structurally conserved regions. These regions can then be used as a framework for the model building. Variable regions, where the known structures may differ in conformation, can be modelled later with other techniques. When only one protein is available as the template structure, usually only the secondary structure elements, known to be the structurally most conserved regions, are used to build the framework. Once the known structures are aligned and the secondary structure conserved regions have been identified, the residues of the conserved framework can be mutated *in silico*. This results in a pre-model where the sequence of the protein of unknown structure is aligned to the conserved framework.

In model building, it should be taken into account that proteins are molecules usually composed of distinct modules, and often different portions of the target sequence may show higher homology with different template proteins, thus different target regions can have different templates.

iii) Addition and optimization of side chains

Once the backbone has been modelled, the side-chains corresponding to the target protein sequence can be added. Mutagenesis *in silico* must take into account preferred side chain torsion angles, close packing, hydrogen bonds, ion pairs and other electrostatic interactions. Rotamer libraries have been constructed (Ponder and Richards, 1987; Dunbrak and Karplus, 1993), reporting the most probable side chain torsion angles depending on the backbone conformation; the most probable rotamer for each residue is added to the model. Parameters such as the solvation energy should be also considered for surface residues. Another typical approach consists of considering the rotamer of the new residue that is most similar to the residue that is substituted.

iv) Loop building

While modelling of secondary structure elements can be very simple, modelling of the loops is a more difficult task, representing a sort of *ab initio* folding problem. However, the knowledge of the surrounding regions may simplify the task. While secondary structure elements are usually conserved in homologous proteins, loop regions may differ since they usually accommodate most insertions, deletions and replacements. Nevertheless, it is now established that loop regions may also have preferred structures rather than a random collection of possible structures, and this makes the modelling approach more feasible.

   If a loop in one of the templates is a good model for a loop present in the target, then the main chain coordinates of that known structure can be used as template backbone of that loop. Rotamer libraries can be used to define other side chain coordinates. When a good model for a loop cannot be found among the known structures, one can search fragment databases for loops in other proteins that may provide a suitable model for the unknown one. Fragments are examined for their ability to fit in the pre-loop and post-loop residues and for their contacts/collisions with other atoms of the model structure. The loop may then undergo conformational searching to identify low energy conformers if desired. Coordinates for side chain atoms in these loop regions may be copied if residues are similar. Alternatively, a search in a side chain rotamer library can be required. CODA (Deane *et al.*, 2001; see Paragraph 10.4) is a loop building procedure available on-line, that combines knowledge-based and *ab initio* approaches.

v) Refinement

The model obtained in the first four steps can be improved using energy minimization procedures as CHARMM (Brooks *et al.*, 1984), AMBER (Weiner and Kollman, 1981), or GROMOS (van Gunsteren *et al.*, 1994). These procedures will not significantly alter the model structure, but usually reduce inconsistencies. If the target shares high sequence identity with the template, it is possible to fix the backbone atoms, considering only the side chain atoms in the calculation. Several rounds of minimization can be performed, but there are evidences that if the procedure is repeated too many times the model becomes less accurate.

vi) Validation

The errors estimate is a necessary but not easy task to validate the model. Different steps can be performed, attempting checks of the fold (i.e. the compatibility of a sequence with the chosen fold) or of stereochemical parameters as bond lengths, bond angles and contacts. Clearly, a positive quality evaluation of the model does not imply that the model is correct, and only an experimental validation could prove it. The homology modelling procedure is generally quite reliable, if the sequence similarity between the target and the templates is not too low. One of the simplest methods to measure and prove the model reliability is to calculate its Ramachandran plot, i.e. a diagram where the backbone psi and phi angles are plotted for all residues in the protein. Only certain regions in the plot are allowed for amino acid residues, and residues falling in not allowed regions are easily detectable (see paragraph 10.1.1.1). The VERIFY3D procedure (Eisenberg *et al.*, 1997) of structure validation is based on the three-dimensional profiles method (Luthy *et al.*, 1992) that considers the statistical preferences of each of the 20 amino acids for the structural environment within the protein, derived from known three-dimensional structures. PROCHECK (Laskowski *et al.*, 1993) checks the stereochemical quality of a protein structure by analyzing its overall and residue-by-residue geometry. The procedure is based on an analysis of parameters such as: phi and psi angles, peptide bond planarity, bond lengths and angles, hydrogen-bond geometry and side-chain conformations. Values found in the proteins of known structures are compared to the model ones. Once irregularities have been checked and resolved with the chosen method, the entire structure may then undergo further refinement.

10.2.2.2 On the web

Although simple in principle, homology modelling theory hides a great complexity. Nevertheless, obtaining a reliable model can, in some instances, be a very simple task. A lot of information about homology modelling can be found easily on the web; some interesting guides for the beginner 'modeller' can also be found. A few resources that can build a model automatically by starting only from the target sequence are available on the web, with user-friendly interfaces and high reliability (they work only with relatively high sequence identities between the target protein and a protein of known structure; see below). For more advanced tasks, powerful software is freely or commercially available and can be downloaded, installed and run locally. Modeller (URL http://guitar.rockefeller.edu/modeller/; Sali and Blundell, 1993) is a powerful and free program that can be executed under Unix or Linux machines; although quite simple to use, this software can be too complex for the average molecular biologist. ModBase (URL http://pipe.rockefeller.edu/modbase/; Sanchez *et al.*, 2000) is a database of three-dimensional protein models calculated with Modeller. Commercial software like Tripos Inc.'s SYBYL® (http://www.tripos.com/software/sybyl.html) can be purchased as part of molecular graphics suites such as InsightII (Dayringer *et al.*, 1986) or as separate packages.

On line tutorials

The Gale Rhodes tutorial is an excellent tutorial centred on the Swiss Model and Swiss PDB Viewer programs, a very clear guide for the beginning user (URL http://www.usm.maine.edu/~rhodes/SPVTut/index.html).

The exhaustive Gert Vriend tutorial was originally written for a course organized by the CBMI (Centre for Molecular and Biomolecular Informatics) in the Nijmegen University (URL http://www.cmbi.kun.nl/gvteach/hommod/index.shtml). In only a few and interesting steps this course, starting from principles of protein structures and structure analysis, takes you to the construction of an energy minimized and validated model.

On-line resources

SWISS-MODEL

SWISS-MODEL (URL http://www.expasy.ch/swissmod/SWISS-MODEL.html) is an automated protein-modelling server developed at GlaxoSmithKline in Geneva, Switzerland; there is no human intervention during the procedure. Some features are designed for the molecular graphics software SwissPDBViewer (see Paragraph 10.1.3). The SWISS-MODEL server accesses the PDB database as a source of structural information and automatically generates protein models for sequences which share significant similarities with at least one protein of known structure. The target sequence can be submitted to the program using a user-friendly form, or a SWISS-PROT accession number can be specified; then the user can choose to let the procedure look for the templates or can specify one or more PDB files as templates. Moreover the user can provide one or more templates to be uploaded from its machine. It is also possible to run a sequence alignment against a sequence database of proteins of known structures, and then select the templates to be used. Whatever the user chooses, the procedure aligns the target to the templates using BLAST P (see chapter 9). The model is built with Promod software (Peitsch, 1996)

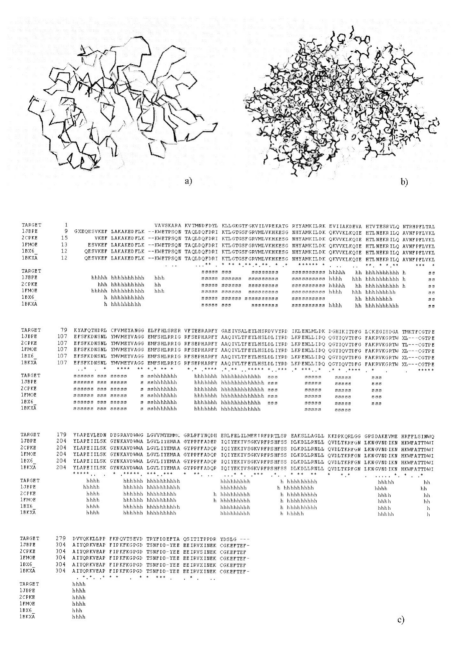

Figure 10.7. Model of the human protein kinase Akt-2. The model of the human protein kinase Akt-2 (SWISS-PROT entry AKT2_HUMAN) obtained by the SWISS-MODEL server. a) superposition of the backbone of the model (colored in black) and the backbone of one of the structures used to build the model framework (the alpha-catalytic subunit of the mouse cAMP-dependent protein kinase, PDB code 1jbp, colored in white); b) superposition of the model and the 1jbp structure. c) sequence alignment calculated by SWISS-MODEL to build the model framework.

and GROMOS96 is used for energy minimization. The resulting model is sent to the user by e-mail, together with information about the modeling procedures, the template structures used and their sequence alignment to the target. The whole procedure usually takes less than one hour. A model obtained with SWISS-MODEL is shown in Figure 10.7.

The reliability of the procedure has been tested by generating models for proteins of known 3D-structure using modelling templates sharing sequence identity in a range from 25 to 95 % with the submitted sequences, collecting over 1200 model-control structure pairs. The degrees of identity between target and template sequences and the relative mean square deviation of the models from their corresponding experimental control structure show that below 30 % sequence identity the accuracy of completely automated protein modelling rapidly degrades. However, 63 % of sequences sharing 40 - 49 % identity with their template yield a model deviating by less than 3 Å from their control structure. This number increases to 79 % for sequence identities ranging from 50 to 59 %. All these models are stored in the SWISS-MODEL repository (URL http://www.expasy.ch/swissmod/ SM_3DCrunch_Search.html), together with a great number of models (more than 60000) obtained by the 3DCrunch project (URL http:// www.expasy.ch/swissmod/SM_3DCrunch.html), whose aims are to model all sequences in the SWISS-PROT and trEMBL databases with significant sequence similarity to proteins with known structure.

CPHmodels

This is another server dedicated to protein structure prediction by homology modelling (URL http://www.cbs.dtu.dk/services/ CPHmodels/) that use a neural network approach improved with sequence profiles (Lund *et al.*, 1997)

FAMS

FAMS (Full Automatic Modeling System, URL http:// physchem.pharm.kitasato-u.ac.jp/FAMS/fams.html) is based on a method for homology modelling consisting of database searches and simulated annealing: the determination of main chain conformations (generated from the weighted average of main chain coordinates in reference proteins) and side-chain conformations (generated by rotamer libraries) are repeated several times in a structural optimization procedure (Ogata and Umeyama, 2000).

10.2.3 Fold Recognition

10.2.3.1 Introduction

When the target does not share significant similarity with any protein of known structure and comparative modelling cannot be applied, other methods can be used to get information about the target structure. Proteins may share very similar folds even when lacking any statistically relevant sequence similarity. The number of folds and topologies that a protein may adopt appears to be limited, so it becomes feasible to construct a library of all the possible folds and evaluate the most 'appropriate' for the sequence of interest. The number of possible protein folds has been estimated to range between 1,000 and 3,000 (Chotia, 1992; Orengo *et al.*, 1994). More than 500 are already known; a few of them are over-represented in the PDB, while others have been seen only once.

The most effective fold recognition technique is known as threading; this method uses a database of known three-dimensional structures to match the target sequences with protein folds. Scoring functions are used to evaluate the fit of a sequence to a given fold, 'mounting' (= threading) the sequence on the structure. The best-scoring fold is proposed to be the one adopted by the target sequence. Several different threading methodologies share this common approach. A number of different scoring methods have been developed to evaluate the fit of a given sequence on a given fold, but generally all of them come directly from observation of the known structures, based on physical rules.

In the simplest form of threading, the target sequence is mounted over a part of a template fold of the same length, without allowing for gaps. Clearly, this assumption may lead to missing the correct fold, but the technique is useful to test and correct the scoring potentials, threading a sequence of known structure on its own fold (Miyazava and Jernigan, 1996). Usually, parts of the target sequence are mounted over parts of the template fold, allowing gaps both in the sequence or the fold; any co-linear alignment between sequence and fold fragments is allowed (Finkelstein, 1997).

Once the sequence is mounted over a fold, energy-like functions are used (Lemer *et al.*, 1995; Jones and Thornton, 1996). These functions may consider only the position of each residue on the fold (its secondary structure, accessibility, angle values, and so on) or may consider the relative position of residue pairs and their interactions. Up to now the problem of finding a reliable set of potentials is not solved, as potentials that work well on some structures may fail on others, and there is not a 'universal' set of potentials.

Even if threading very often proves reasonably efficient in the identification of the correct fold for a target sequence, it sometimes shows inconsistencies in the proposed sequence-structure alignment. The problem may depend on the differences between similar folds when the sequence share a weak sequence similarity with the sequence of the selected original fold: when the rmsd between the folds of two distant homologous proteins is greater than 2Å, the sequence-structure alignment of a sequence with the fold of the other is less than 50% correct (Wilmanns and Eisenberg, 1995; Bryant, 1996).

More recently other approaches were successfully developed; they rely on secondary structure prediction and multiple sequence alignment. These methods compare the similarity of a target sequence to a template structure using profiles derived from sequence, solvent accessibility, and secondary structure information, and they now appear to be at least equivalent in performance to threading methods. Koretke *et al.* (1999) used PSI-BLAST and Hidden Markov Models (see Chapter 9) to obtain a sequence profile; the alignment was used to obtain a secondary structure prediction with the Jpred program (see Parargraph 10.2.1). The Jpred prediction provides the input for the MAP program (Russell *et al.*, 1996) that aligns the predicted secondary structure elements to those observed in known structures. Alignments are then filtered according to some structural rules, and then ranked.

Jones (1999b) proposed a method that takes advantage of both threading and sequence profile procedures. The method, called GenTHREADER (which is part of the PSIpred structure prediction server, URL http://insulin.brunel.ac.uk/psipred/; McGuffin *et al.*, 2000), generates sequence profile-based alignments, which are evaluated by threading techniques; then a neural network is used to assess each threaded model (Jones *et al.*, 1999).

The 3D-PSSM (three-dimensional Position-Specific Scoring Matrix) method (Kelley *et al.*, 2000) uses structural alignments of homologous proteins, based on the SCOP classification (see Paragraph 10.1.1.2), to obtain structurally equivalent residues. The method considers a representative member of each SCOP superfamily, from which a sequence profile is obtained. Using the SAP program (Orengo *et al.*, 1992; Taylor and Orengo, 1989), a three-dimensional structural superposition is made among all the proteins belonging to the same SCOP superfamily (this is made only for the structures showing an rmsd less than 6Å versus the representative structure). The goal is to obtain a sort of three-dimensional matrix in which each position contains sequence and structural information (to this purpose secondary structure and solvation information are also considered). All these matrices compose a library that can be scanned using a query sequence using an algorithm derived from the fold recognition program FOLDFIT (Russell *et al.*, 1998). The score for a match between a residue in the probe and a residue in the library sequence is calculated as the sum of the secondary structure, solvation potential and profiles scores. The procedure is freely available on-line (see Paragraph 10.2.3.2).

10.2.3.2 On the web

Several on-line resources for threading became available in the last years; some of them can be used directly on remote servers, while other programs can be downloaded, installed and run locally, as the PROSPECT program (URL http://compbio.ornl.gov/structure/prospect/, Xu and Xu, 2000).

3D-PSSM

3D-PSSM (URL http://www.bmm.icnet.uk/~3dpssm/) is a method for protein fold recognition which uses 1D and 3D sequence profiles coupled with secondary structure and solvation potential information. A number of scoring components are used: 1D-PSSMs (sequence profiles built from relatively close homologues), 3D-PSSMs (more general profiles containing more remote homologues), matching of secondary structure elements, and propensities of the residues in the target sequence to occupy varying levels of solvent accessibility. The query can be a single sequence but also a multiple alignment or a sequence profile. Results (alignments for the 20 most probable matches in the library and a PSI-Pred secondary structure prediction) are rapidly sent by e-mail (usually less than half an hour), while a more informative HTML version of the results is available at a link provided to the user.

Bioinbgu

The Bioinbgu bioinformatic server (URL http://www.cs.bgu.ac.il/~bioinbgu/) provides fold-recognition services based on sequence-derived properties, with a methodology called by the author 'hybrid' fold recognition (Fischer, 2000). The whole procedure is accomplished by submitting the target sequence to five different evaluation programs, using the target sequence itself, a PSI-BLAST multiple alignment or a profile, together with a PHD secondary structure prediction, to search through a fold database comprising 3000 folds in the latest release. Each program provides an output that is evaluated by a 'consensus' method that provides the final fold prediction.

FUGUE

FUGUE (Shi *et al.*, 2001) is a program designed for recognizing distant protein homologues by sequence-structure comparison (URL http://www-cryst.bioc.cam.ac.uk/fugue/), combining information from both multiple sequences and multiple structures. It aligns multiple sequences against multiple structures, enhancing in this way the recognition of conserved features, and it is based on the Homologous Structure Alignment Database (HOMSTRAD, URL http://www-cryst.bioc.cam.ac.uk/~homstrad/; Mizuguchi *et al.*, 1998). Alignments take into account parameters of local structural environment (i.e. secondary structure, solvent accessibility, hydrogen bonding status).

10.2.4 *Ab initio* Prediction

10.2.4.1 Introduction

Attempts to predict tertiary structure in the absence of any homology to a known structure by trying to simulate the natural protein folding process are known as *ab initio* procedures. *Ab initio* structure prediction procedures can be based on different approaches and methodologies, and can be classified according to the dimensionality of the resulting information (Jones, 1997): 0D (i.e. prediction of the fold class), 1D (prediction of secondary structure or accessibility), 2D (inter-molecular contact prediction, such as correlated mutation method), 3D (protein folding simulations). While secondary prediction methods have been described in Paragraph 10.2.1 of this Chapter, the first class of *ab initio* methods (0D methods) are represented by early studies regarding the possibility of predicting the fold class using only information derived from the amino acidic composition (Nakashima *et al.*, 1986; Eisenhaber *et al.*, 1996).

The prediction of the inter-molecular contacts of a protein structure (i.e. the prediction of which residues will be in contact in the folded protein) can be used as a guide to obtain a folding simulation; contacts can be predicted through the analysis of the correlated sequence changes between different positions in a multiple alignment (Gobel *et al.*, 1994; Thomas *et al.*, 1997): if a mutation is unfavourable, natural selection leads to its elimination but nevertheless, a compensatory mutation might occur, thus preserving the structure and the function of the protein. The contact map prediction can be improved with other data derived from multiple sequence alignments, such as sequence conservation, sequence separation along the chain, alignment stability, family size, residue-specific contact occupancy and formation of contact networks (Olmea and Valencia, 1997).

The *ab initio* prediction of protein tertiary structure can be further divided into two groups: the first tries to build and evaluate with energy functions and molecular dynamics tools low-resolution representations of the protein structure; the other group tries to pack together the predicted secondary structure elements in the most probable way. The former class of methods is based on the generation of protein chain conformations, each one evaluated by a potential function. To overcome problems due to the great intrinsic complexity of the folding process, many *ab initio* prediction methods aim at reducing the protein structures into low-complexity models. Some approaches rely on lattice models, converting the chain conformation into a series of points in a lattice (Hinds and Levitt, 1994; Skolnick and Kolinski, 1991). These procedures offer the advantage of computational and analytical simplicity, although having obvious limitations due to the simplified structure representation (Park and Levitt, 1995). Other attempts to reduce protein structure complexity are based on the limitation of the side chains degrees of freedom, together with fixing the psi and phi angle values to a given set of possible pairs (Park and Levitt, 1995; Simons *et al.*, 1997). Generally, to obtain all the putative structures to be evaluated, a sort of Monte Carlo optimization or a genetic algorithm is used (Pedersen and Moulth, 1996; Rabow and Scheraga, 1996; Ortiz *et al.*, 1998). Once a desired structure representation is chosen, an energy function that simulates the forces that are responsible for the protein structure, as solvation and intra-chain bonds, must be applied. These scoring functions can be derived from databases of known structures. Although threading potentials could be used, more effective functions for *ab initio* predictions consider additional terms such as steric hindrance, hydrogen bonding and general chain compactness. The low-complexity approach, though effective, is not expected to produce models with high resolution, never better than 3 - 7 Å (Bonneau and Baker, 2001).

Given a secondary structure prediction, secondary structure elements can be assembled constructing plausible tertiary structures that could be evaluated and sorted. Features such as the association of ß strands into sheets can be considered in a scoring function: the predicted protein tertiary structure can be evaluated calculating the probability of all the different strand arrangements or other kind of secondary structure elements interactions, such as helix-strand interactions (Cohen *et al.*, 1980; Cohen *et al.*, 1982; Reva and Finkelstein, 1996; Simons *et al.*, 1999).

The Rosetta method (Simons *et al.*, 1999), one of the most effective *ab initio* procedures up to date, uses a Monte Carlo search of the possible combinations of local structure, evaluated with a scoring function that takes into account factors as hydrophobicity, residue pairs interactions and ß strands pairing.

10.2.4.2 On the web

A few servers dedicated to *ab initio* prediction are now available, exploiting different approaches. Most of them recommend careful consideration of the results, and to use their predictions only if more reliable approaches, such as homology modelling, cannot be applied. The CORNETT (URL http://prion.biocomp.unibo.it/cornet.html) and the PDG-contact-pred server (URL http://montblanc.cnb.uam.es/ cnb_pred/cgi-bin/contact_pred_cgi) are dedicated to intramolecular contact predictions using the correlated mutations approach. The I-sites/Rosetta server (URL http://honduras.bio.rpi.edu/~isites/ ISL_rosetta.html) uses the Rosetta method, together with a hidden Markov model approach (HMMSTR, Bystroff *et al*, 2000), and is considered one of the most reliable. Other servers are reported in section 10.4.

10.2.5 Evaluation of structure prediction methods

Every two years, starting from 1994, the protein prediction community has been involved in a competition to assess the current state of the art in the field, known as CASP (Critical Assessment of Protein Structure Prediction). Each scientific team can try to predict the structure of a number of proteins with known but not yet available structures; the submitted predictions are evaluated by a group of experts with different evaluation methods. Predictions were divided into three categories: homology modelling, fold recognition and *ab initio* prediction methods. The results, reported at the Protein Structure Prediction Center web site (URL http://predictioncenter.llnl.gov), are then analyzed and commented on by experts in the field, with particular attention to the improvement over previous editions (Jones, 1997; Moult, 1999; Sternberg *et al.*, 1999). There has been a clear improvement over the years in protein structure prediction, which has involved all the three major categories. Great progress has occurred in homology modelling, as a consequence of improvement in sequence alignment methods, but the largest development involved the *ab initio* prediction category. This is now an effective tool, while only a few years ago it was considered only as a difficult, intellectual challenge.

Together with improvements in protein structure prediction reliability, several methods have become available on-line, and predictions can now be obtained through easy-to-use web interfaces. In 1998 Daniel Fischer of the Ben Gurion University (Israel) organized the CAFASP (Critical Assessment of Fully Automated Structure Prediction) with the intent of evaluating the capability of different prediction programs without any human intervention. The participants in the CAFASP competitions are Internet servers and programs: target proteins taken from the CASP competitions are submitted and the resulting models evaluated. The first CAFASP experiment showed that automatic procedures are able to produce reasonable predictions, although human intervention can improve the results (Fischer *et al.*, 1999). CAFASP2 was held in parallel with CASP4, and several automatic procedures were involved.

Large-scale evaluations of automated protein structure prediction methods are ongoing. Eva is a web server (URL http://pipe.rockefeller.edu/~eva/) that accomplishes an automated evaluation of protein structure prediction servers. This is made by taking new structures from the PDB, submitting them to various participating modelling servers and evaluating the predictions. Different parameters are tested (correctness of the fold, correctness of the alignment, stereochemical quality, accuracy of core and non-core regions, overall quality). The evaluated servers are classified in categories (homology modelling, threading, secondary structure prediction, inter-molecular contacts); each week more than 20 targets corresponding to PDB structures are submitted to the servers. Results are still not discussed but can be downloaded. The LiveBench Project (URL http://BioInfo.PL/LiveBench; Bujnicki *et al.*, 2001) is another continuous evaluation program: every week, new PDB proteins are submitted to participating fold recognition servers. The results are collected and evaluated using automated model assessment programs. Up to now, nine fold recognition servers are considered.

10.3 Transmembrane topology prediction

10.3.1 Introduction

Several proteins of primary biological importance, such as receptors or channels, cross the membrane lipid bilayer with part of their structure. It has been estimated that in the yeast genome 35 - 40% of the genes code for membrane proteins (Goffeau *et al.*, 1993). Statistical analysis of integral membrane proteins of the helix-bundle class (the most abundant) from eubacterial, archaean, and eukaryotic genomes showed that 20 - 30 % of ORFs are predicted to code for membrane proteins (Wallin and von Heijne, 1998; Stevens and Arkin, 2000). The general shape of membrane proteins presents many transmembrane segments joined by short loops, or a few transmembrane portions with large cytoplasmic or external domains. Transmembrane protein topology generally can adopt two different architectures: the helix bundle and the beta-barrel (Figure 10.8). Hydrophobic residues organized in alpha-helices compose the former, while the latter is a cylindrical structure composed by antiparallel beta-sheets. The presence of a protein segment in a lipidic membrane implies that charged or highly polar uncharged groups must be shielded from the oil-like hydrocarbon interior of bilayer membranes, and this step has a very high thermodynamic cost. For this reason charged or polar residues are rarely found in transmembrane regions, but the CO and NH groups of the peptide bond itself have polar characteristics. Thus, alpha-helices, in which all these groups are involved in H-bonds are the structures that can be transferred to the membrane with the lowest energetic cost. A beta-sheet with a close structure, as a beta-barrel, also has all main-chain polar groups involved in H-bonds and can then represent an intramembranous structure.

These structures usually have distinctive patterns of hydrophilic

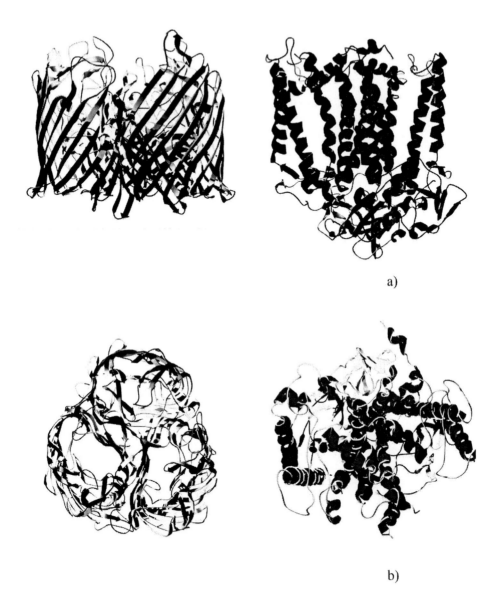

a)

b)

Figure 10.8. Schematic representation of the *Rhodobacter sphaeroides* photosynthetic reaction center and the *Eschericha coli* maltoporin. Two membrane proteins, the beta-barrel *Rhodobacter sphaeroides* photosynthetic reaction center (PDB code 1aig) on the left side and the helix bundle *Eschericha coli* maltoporin (PDB code 1af6) on the right, are shown in a ribbon schematic, obtained with the SwissPDBViewer molecular graphics software. a) side view; b) upper view.

Table 10.2. The Wimley-White hydropathy scale. The Wimley-White scale reports the experimentally determined free energies (ΔG) of transfer from water to POPC (palmitoyl-oleoyl phosphatidylcholine) interface (*wif*) and to *n*-octanol (*woct*). It is used by the MPEx transmembrane helix prediction procedure.

Aminoacid	Interface Scale ΔG_{wif} (kcal/mol)	Octanol Scale ΔG_{woct} (kcal/mol)	Octanol - Interface Scale
Ala	0.17	0.50	0.33
Arg +	0.81	1.81	1.00
Asn	0.42	0.85	0.43
Asp -	1.23	3.64	2.41
Asp 0	-0.07	0.43	0.50
Cys	-0.24	-0.02	0.22
Gln	0.58	0.77	0.19
Glu -	2.02	3.63	1.61
Glu 0	-0.01	0.11	0.12
Gly	0.01	1.15	1.14
His +	0.96	2.33	1.37
His 0	0.17	0.11	-0.06
Ile	-0.31	-1.12	-0.81
Leu	-0.56	-1.25	-0.69
Lys +	0.99	2.80	1.81
Met	-0.23	-0.67	-0.44
Phe	-1.13	-1.71	-0.58
Pro	0.45	0.14	-0.31
Ser	0.13	0.46	0.33
Thr	0.14	0.25	0.11
Trp	-1.85	-2.09	-0.24
Tyr	-0.94	-0.71	0.23
Val	0.07	-0.46	-0.53

and hydrophobic residues in their primary sequence that correspond to loops or intramembranous regions, respectively. An intramembrane helix can appear in the sequence as a stretch of hydrophobic residues, while a membrane beta-strand has an amphipathic nature. It is thought that a simple alternate pattern of hydrophobic and hydrophilic residues is not sufficient for transmembrane region identification, as the amphipathicity may be more complex.

These protein structures are difficult to determine with current techniques. For this reason, prediction methods to infer transmembrane

```
# Sequence Length: 281
# Sequence Number of predicted TMHs:  5
# Sequence Exp number of AAs in TMHs: 113.10745
# Sequence Exp number, first 60 AAs: 24.19884
# Sequence Total prob of N-in:        0.93939
# Sequence POSSIBLE N-term signal sequence
Sequence    TMHMM2.0    inside       1     28
Sequence    TMHMM2.0    TMhelix     29     51
Sequence    TMHMM2.0    outside     52     82
Sequence    TMHMM2.0    TMhelix     83    105
Sequence    TMHMM2.0    inside     106    111
Sequence    TMHMM2.0    TMhelix    112    134
Sequence    TMHMM2.0    outside    135    174
Sequence    TMHMM2.0    TMhelix    175    197
Sequence    TMHMM2.0    inside     198    233
Sequence    TMHMM2.0    TMhelix    234    256
Sequence    TMHMM2.0    outside    257    281
```

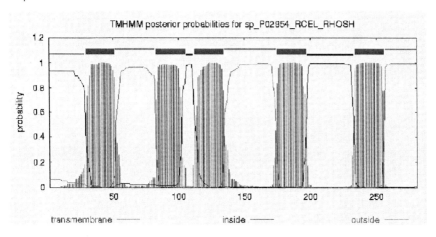

Figure 10.9. Transmembrane helices prediction for the *Escherichia coli* maltoporin. The sequence of the *Escherichia coli* maltoporin (SWISS-PROT entry RCEL_RHOSH, PDB code 1af6) has been submitted to the TMHMM prediction procedure. The five transmembrane helices are correctly predicted.

segments in the protein sequence have great relevance and are widely used. An approach to inferring the position of transmembrane regions in a protein is the building of a hydrophobicity plot, which is obtained by plotting residue hydrophobicity as a function of sequence position (Figure 10.9). Several hydrophobicity scales, composed of experimentally determined transfer free energies for each amino acid, have been determined (Kyte and Doolittle, 1982; Engelman *et al.*, 1986; Wimley and White, 1996). Plots are scanned through 15 - 25 residue windows, looking for stretches of hydrophobic amino acids.

In Table 10.2 the Wimley-White scales are reported, in which the free energies (ΔG) of transfer from water to POPC (palmitoyl-oleoyl phosphatidylcholine) interface (*wif*) and to *n*-octanol (*woct*) are shown.

Another clue to the presence of intramembrane helices can be the abundance of arginines and lysines in the predicted cytoplasmic portions (the so-called 'positive inside rule'; von Heijne and Gavel, 1988): all acceptable predicted topologies should have loops with a high fraction of positively charged residues on the cytoplasmic side.

Several techniques are available to identify protein transmembrane helical regions, and most of them are based on hydropathy scales. The use of neural networks (Rost *et al.*, 1996) or weight matrices (Edelman, 1993) can improve the prediction reliability. Most widely used methods are presented in the following section. In contrast, only few methods have been developed for transmembrane beta-strand prediction, and they are often based mainly on multiple sequence alignments, physico-chemical properties of amino acid composition and analysis of amphipathicity (Jeanteur *et al.*, 1991; Welte *et al.*, 1991; Schirmer and Cowan, 1993; Gromiha *et al.*, 1997). Diederichs *et al.* (1998) proposed an approach based on a neural network, trained on seven porin structures, solved at high resolution.

10.3.2 On the web

MPEx

The Membrane Protein Explorer (MPEx) is a web server (URL http://blanco.biomol.uci.edu/mpex/index.html#FeaturesGlance) that uses a Java applet to get hydropathy plots according to the Wimley-White hydropathy scale (Wimley and White, 1996), using an algorithm based upon a 19-residue sliding window to select the regions most likely to be transmembrane segments. MPEx is also available as a stand-alone application installable on the most used operating systems.

Linked to the MPEx server is the MPtopo database (URL http://blanco.biomol.uci.edu/mptopo/index.html; Jayasinghe *et al.*, 2001), which is an annotated collection of proteins containing experimentally verified transmembrane segments (by means of crystallography, gene fusion, proteolytic degradation, and others), containing about one hundred proteins and updated regularly.

TMHMM

The TMHMM method (Krogh *et al.*, 2001), which is available on-line at http://www.cbs.dtu.dk/services/TMHMM-2.0/, relies on the use of hidden Markov models to find the most probable topology using a set of rules, i.e. the probability of each of the 20 amino acids to be in different states (such as intramembranous, cytoplasmic, cytoplasmic close to the membrane, and so on) and the length distribution of transmembrane helices, based on a training set of 160 well-characterized membrane proteins. The procedure is able to correctly predict about 78% of the membrane helices in the training set; the percentage raises to 84% if the prediction is correct but the topology is inverted (helices are correctly inferred, but the cytoplasmic portion is predicted to be external, and the external cytoplasmic). The TMHMM procedure can be used to discriminate between membrane and soluble proteins, and it has been applied to several fully sequenced genomes, showing that the helix bundle membrane proteins ratio varies between 20 - 30%, a result similar to previous analyses, with few exceptions (the *Plasmodium falciparum* genome seems to have a higher fraction, about 40%).

Sequences to be analyzed can be submitted to the web server interface in FASTA format. The prediction output reports the number of predicted transmembrane helices, the length and the sequence range of each helix; a graphical representation can be obtained (Figure 10.9).

A systematic evaluation of the most widely used protein transmembrane regions prediction methods (Möller *et al.*, 2001) showed that TMHMM is the procedure with the best overall performance.

HMMTOP

The HMMTOP method (Hidden Markov Model for Topology Prediction, which can be used on-line at http://www.enzim.hu/hmmtop/

index.html), uses a hidden Markov model to identify differences in the amino acid distribution in various structural parts of a protein training set (Tusnady and Simon, 1998). A unique feature of the procedure is that if information about the localization of portions of the query protein is known, then the prediction can be performed using them as restrictions, improving the overall reliability.

MEMSAT

MEMSAT is an all-helical membrane protein prediction program that is based on the recognition of topological models derived from well-characterized membrane proteins (Jones *et al.*, 1994). The procedure can be accessed at http://www.biology.ucsd.edu/~yzhai/memsat.html, while its latest version (MEMSAT 2) is part of the PSIpred protein prediction server (see the Links section). Tested on a set of 86 transmembrane proteins of known topology, MEMSAT 2 has an accuracy of 92%.

PHDhtm

The PHDhtm procedure is part of the PredictProtein server (see 10.4 paragraph); it performs transmembrane helix predictions using a neural network system and multiple sequence alignments (Rost *et al.*, 1996). The 'positive inside' rule is applied to improve the accuracy. On a test set made by 131 transmembrane proteins the procedure had 86% prediction accuracy.

10.4 Links

10.4.1 Protein structure

CATH

URL   http://www.biochem.ucl.ac.uk/bsm/cath_new/index.html
The CATH database contains a hierarchical classification of protein domains. The acronym derives from the four major organization levels: class (C), architecture (A), topology (T) and homologous superfamilies (H). The database can be accessed through a key-word search or browsed starting from the highest level (the class level, divided in mainly alpha structures, mainly beta, mixed alpha-beta and structures with few secondary structure elements). Raw files containing the whole database can be downloaded from an ftp site.

DALI

URL   http://www2.embl-ebi.ac.uk/dali/
The DALI server is a tool for comparing protein 3D structures (Holm and Sander, 1997). The user can submit the coordinates of a query structure (by e-mail or through a web form), and choose to compare the query against the Protein Data Bank or attempt a pair-wise comparison against a second submitted structure. A multiple (or a pair-wise) structural alignment is mailed back to the user. The DALI procedure has been used to build the structure classification database DALI Domain Dictionary (URL http://www2.embl-ebi.ac.uk/dali/domain/).

DSSP

URL   http://www.cmbi.kun.nl/swift/dssp/
The DSSP database contains information about secondary structures, hydrogen bonds and solvent accessibility for each PDB file derived with the DSSP program (Kabsch and Sander, 1983).

FSSP

URL   http://www2.embl-ebi.ac.uk/dali/fssp/
The FSSP database (Fold classification based on Structure-Structure alignment of Proteins) contains structural protein alignments obtained automatically with the DALI program on a non-redundant set of proteins. A hierarchical clustering method is used to construct a tree based on the structural similarities from the all-against-all comparison; the user may access the database from the top of this tree or through a key-word search.

HSSP

URL   http://www.cmbi.kun.nl/swift/hssp/
The Homology-derived Structures of Proteins database contains the sequence alignments of all the PDB proteins with all the PDB and SWISS-PROT proteins sharing at least 30% sequence identity with a protein of known structure.

PDB

URL   http://www.rcsb.org/pdb/
The Protein Data Bank is a protein, nucleic acid and carbohydrate structure database, determined with X-ray diffraction or NMR (a few theoretical models are stored, too). New structures are submitted by research groups, evaluated and annotated by the Research Collaboratory for Structural Bioinformatics (RCSB) staff. The DB is updated weekly; at 4th September 2001 it contained 15,953 structures. Different query interfaces allow the user to find structures of interest, which can be viewed through a browser or downloaded in different formats.

PDBSum

URL   http://www.biochem.ucl.ac.uk/bsm/pdbsum/
The PDBsum database contains summary information and derived data on each structure of the PDB. Data include PROCHECK, PROMOTIF and LIGPLOT analyses, together with links to several databases.

SARF

URL   http://www-lmmb.ncifcrf.gov/~nicka/sarf2.html
The SARF2 (Spatial ARrangement of backbone Fragments) algorithm (Alexandrov, 1996) tries to find the largest set of secondary structure elements of two structures that can be superposed, minimizing the root mean square deviation.

SCOP

URL   http://scop.mrc-lmb.cam.ac.uk/scop/
The Structural Classification Of Proteins database is a classification of all the proteins with known structure based on structural and evolutionary analysis. The classification is made manually by visual inspection and comparison of structures, but with the assistance of automated tools. The classification is hierarchical, with different levels that reflect the structural similarities and the evolutionary relationships between the proteins. The database can be scanned entering at the top hierarchical level or through a keyword search.

TOP

URL   http://bioinfo1.mbfys.lu.se/TOP/webtop.html
The TOP program (Lu, 2000) compares protein structures looking for common substructures. The algorithm represents each structure as a group of secondary structure elements, and systematically searches all the possible superpositions of these elements. Once a couple of

elements in the two structures can be superposed with a root mean square deviation lower than a cut-off value, the program will search whether more secondary structure elements can fit by the same superposition operation. The software can be freely downloaded and installed locally.

10.4.2 Molecular graphics tools

Chime

URL   http://www.mdlchime.com/chime/
Chime is a plug-in for Netscape Communicator and Microsoft Internet Explorer browsers from MDL Information Systems, Inc., that interactively displays 2D and 3D molecules directly in Web pages.

Cosmo Player

URL   http://www.cai.com/cosmo/
Cosmo Player from Cosmo Software is a VRML plug-in for Netscape and Internet Explorer, which allows the visualization and manipulation of molecules directly through the browser, and that can be downloaded freely.

RasMol

URL   http://www.bernstein-plus-sons.com/software/rasmol/
This is the home page for the popular free software for molecular graphics, from which it is possible to download the program for MacOS, Windows and Unix, and to access an exhaustive manual.

Swiss PDB Viewer

URL   http://www.expasy.ch/spdbv/
The Swiss PDB Viewer is one of the most complete molecular graphics

packages, and it is possible to download it for free from this site for the most popular operating systems. A very nice tutorial can be found, to easily learn all the basic features and some advanced ones (such as how to change residues, simulating mutations).

10.4.3 Secondary structure prediction

Jpred2

URL   http://jura.ebi.ac.uk:8888/

JPred is a web server that allows the user to submit a sequence or a multiple alignment and get the prediction of several secondary prediction methods, chosen from among the most accurate ones. Results from each method are shown in a web page, together with a consensus prediction that takes into account each single result. This combination has been shown to be very successful.

Pred2ary

URL   http://www.cmpharm.ucsf.edu/~jmc/pred2ary/

This method (Chandonia and Karplus, 1999) for secondary structure prediction uses neural networks that analyze multiple alignments obtained by PSI-BLAST. Probabilities for H (alpha-helix), E (beta-strand), and C (everything else) are predicted at every position in the sequence. The program can be used on-line through a web browser, or can be downloaded and executed locally.

PREDATOR

URL   http://www.embl-heidelberg.de/argos/predator/run_predator.html

PREDATOR (Frishman and Argos, 1996; Frishman and Argos, 1997) is a secondary structure prediction program which uses pair-wise local alignments between the target sequence and a set of related sequences, rather than multiple alignments, reaching a 75% accuracy. The procedure can be used through the on-line server (the results are e-mailed to the user), while the source code for UNIX and DOS can be downloaded by anonymous FTP from ftp.ebi.ac.uk.

PSIpred

URL   http://insulin.brunel.ac.uk/psipred/
The PSIPRED protein prediction server (McGuffin *et al.*, 2000) allows the user to perform a secondary structure prediction (with the PSIPRED method), a transmembrane topology prediction (with MEMSAT 2) and a fold recognition procedure (with GenTHREADER); results are sent by e-mail and are available with a graphic representation on web page.

SAM-T99

URL   http://www.cse.ucsc.edu/research/compbio/HMM-apps/
T99-query.html
The Sequence Alignment and Modeling system (SAM; Krogh *et al.*, 1994) uses linear Hidden Markov Models (HMM) to build multiple alignments, which can be used for secondary structure prediction or fold recognition by searching the PDB for similar proteins; a library of HMMs built from proteins with known structures is used to score the target sequence.

SSpro

URL   http://promoter.ics.uci.edu/BRNN-PRED/
This is a server for automatic secondary structure prediction with very high reliability. The procedure utilizes bi-directional neural networks to analyze profiles generated with a PSI-BLAST derived algorithm. Usually results are sent by e-mail.

10.4.4 Homology Modeling

Amber Home page

URL   http://www.amber.ucsf.edu/amber/
AMBER is a molecular mechanical force field for the simulation of biomolecules (which is in general use in a variety of simulation programs); the license is not free. The AMBER server (URL http://narfi.compchem.ucsf.edu/) is a web-based resource that allows you to carry out molecular mechanics minimizations and energy evaluations of a biomolecule using AMBER 6.0; you can upload your PDB file to the server and receive the results by e-mail.

CHARMM

URL   http://www.scripps.edu/brooks/charmm_docs/charmm.html
CHARMM (Chemistry at HARvard Macromolecular Mechanics) is a program for macromolecular dynamics and mechanics. It can be used for energy minimization, normal modes and crystal optimizations. Linux and Unix versions are freeware.

CMBI Homology Modeling course

URL   http://www.cmbi.kun.nl/gvteach/hommod/index.shtml
This material derives from a course organized by the Centre for Molecular and Biomolecular Bioinformatics. A number of clear lessons teach the student how to construct a reliable model. Really useful and interesting for all beginners.

CODA

URL   http://www-cryst.bioc.cam.ac.uk/~charlotte/Coda/
CODA (Deane and Blundell, 2001) is a tool for the prediction of

structurally variable protein fragments, ranging around 3 -8 residues. It is based on two programs: FREAD (a knowledge-based method based on a database of PDB fragments) and PETRA (an *ab initio* method based on a database of computer generated conformers).

CPHmodels

URL   http://www.cbs.dtu.dk/services/CPHmodels/
A neural network based homology modelling structure prediction server (Lund *et al.*, 1997), with a very simple interface: target sequence can be pasted in a window, and results are sent by e-mail.

FAMS Modeling service

URL   http://physchem.pharm.kitasato-u.ac.jp/FAMS/fams.html
FAMS (Full Automatic Modelling System) is based on a homology modelling method, which includes a structural optimization process. The target sequence can be submitted in a free form, and the results are sent by e-mail.

Gromos Home page

URL   http://igc.ethz.ch/gromos/welcome.html
GROMOS is a general-purpose molecular dynamics computer simulation package which can be used to: simulation of arbitrary molecules in solution or crystalline state by the method of molecular dynamics (MD), stochastic dynamics (SD) or path-integral method; energy minimization of arbitrary molecules; analysis of conformations obtained by experiment or by computer simulation. The software license is not free, with a higher amount for commercial organization than for academics.

ModBase

URL   http://pipe.rockefeller.edu/modbase/

ModBase is a database of three-dimensional protein models calculated by the MODELLER software. It contains 243,410 models for domains in 159,686 proteins and 371,816 PSI-BLAST fold assignments for domains in 197,999 proteins (updated on July 2000).

Modeller

URL   http://guitar.rockefeller.edu/modeller/

MODELLER is a program for homology modelling of protein structure, which can also carry out multiple comparisons of protein sequences and/or structures, clustering of proteins, and searching of sequence databases. The software is available for academic users by anonymous ftp (ftp://guitar.rockefeller.edu/pub/modeller/), and can be downloaded and installed on Unix machines.

Model Validation server

URL   http://www.cmbi.kun.nl/swift/servers/modcheck-submit.html

This is a tool to evaluate the correctness of a model with the WHAT_IF procedure. The user can submit a model of a protein structure for validation, and the produced report with all the relevant checks for the model can be accessed in a web page.

Molecular Modelling for beginners

URL   http://www.usm.maine.edu/~rhodes/SPVTut/index.html

This excellent tutorial provides an introduction to macromolecular modelling with Swiss PDB Viewer, including a review of many basic concepts in protein structure.

PROCHECK

URL   http://www.biochem.ucl.ac.uk/~roman/procheck/procheck.html

PROCHECK is useful software to check the stereochemical quality of a protein structure or the reliability of a model, highlighting regions whose geometry is unusual. The source code and documentation are available by anonymous ftp after signing a confidentiality agreement. As you can read in the operating manual, 'The aim of PROCHECK is to assess how normal, or conversely how unusual, the geometry of the residues in a given protein structure is, as compared with stereochemical parameters derived from well-refined, high-resolution structures. Unusual regions highlighted by PROCHECK are not necessarily errors as such, but may be unusual features for which there is a reasonable explanation (*e.g.* distortions due to ligand-binding in the protein's active site). Nevertheless they are regions that should be checked carefully.'

SWISS-MODEL

URL   http://www.expasy.ch/swissmod/SWISS-MODEL.html

SWISS-MODEL is an automated protein modelling server. The user can paste the target sequence in a form, and the server will automatically execute the whole procedure, sending by e-mail to the user the final 3D model. It is very fast and reliable. For other details see Section 10.1.3.

SWISS-MODEL repository

URL   http://www.expasy.ch/swissmod/SM_3DCrunch_Search.html

This database stores more than 60000 models obtained with the SWISS-MODEL automated comparative modeling procedure. The majority of the models were built during the 3DCrunch project, where models for every protein in the SWISS-PROT and trEMBL database were built by comparative methods (using SWISS-MODEL).

Furthermore, another aim of the project is to use a fold recognition algorithm (FoldFit) to examine sequences of bacterial origin, which are not significantly similar to proteins of known structure, and can therefore not be modelled. Taken together, these approaches will yield structural models for all sequences with clear similarities to proteins of known 3D structure and a suggested fold class for all bacterial sequences.

VERIFY3D

URL   http://www.doe-mbi.ucla.edu/Services/Verify_3D/
The Verify3D Structure Evaluation server is a tool designed to help in the refinement of crystallographic structures, submitted in PDB format by uploading them from the user computer.

WHAT IF

URL   http://www.cmbi.kun.nl:1100/WIWWWI/
This server allows to build, refine and evaluate a protein structure model and to compare the model to a protein with known structure; moreover, the server offers several additional features, as the possibility to calculate a Ramachandran plot of a given structure or the solvent accessible surface.

10.4.5 Fold recognition

123D+

URL   http://www-lmmb.ncifcrf.gov/~nicka/123D.html
123D+ is a web server hosting a threading program which combines substitution matrices, secondary structure prediction, and contact capacity potentials.

3D-PSSM

URL   http://www.bmm.icnet.uk/~3dpssm/
The 3D-PSSM server is designed to take a protein sequence and predict its 3-dimensional structure and its probable function. Each sequence is 'threaded' and scored versus a library of known protein structures. A variety of scoring components are employed: 1D-PSSMs (sequence profiles built from relatively close homologues), 3D-PSSMs (more general profiles containing more remote homologues), matching of secondary structure elements, and propensities of the residues in the query sequence for different degrees of solvent accessibility.

Bioinbgu

URL   http://www.cs.bgu.ac.il/~bioinbgu/
Bioinbgu is the bioinformatics homepage of the Computer Science Department of the Ben Gurion University, and provides a fully automated fold recognition procedure based on improved multiple sequence alignments and secondary structure prediction, with a methodology called 'hybrid' fold recognition (Fischer, 2000). The user can submit a query by e-mail or fill an online form. The results are temporarily stored in a web page.

FFAS

URL   http://bioinformatics.ljcrf.edu/FFAS/
FFAS (Fold and Function Assignment System; Rychlewski *et al.*, 2000) is a profile-profile alignment method, which uses PSI-BLAST to collect homologues of the query sequence, then the selected sequences are converted into a profile that is aligned against a profile database.

FORESST

URL   http://abs.cit.nih.gov/foresst/foresst.html
FORESST (FOld REcognition from Secondary Structures, Di

Francesco *et al.*, 1999) is a database of hidden Markov models representing the secondary structures of protein structural domains. The models are based on structural families in the CATH database. A search engine allows the comparison of a predicted secondary structure query sequence against the database, and generates a prediction of the possible fold.

FUGUE

URL   http://www-cryst.bioc.cam.ac.uk/fugue/
FUGUE is a procedure for the identification of protein distant homologues that is based on sequence and structure multiple alignments. Structural alignment information is stored in the Homologous Structure Alignment Database (HOMSTRAD, URL http://www-cryst.bioc.cam.ac.uk/~homstrad/). Each alignment in HOMSTRAD is converted into a scoring template, then the program FUGUE compares a sequence or a multiple alignment against the profiles library. The query is submitted using a web form; results are temporarily stored in a web page.

LOOPP

URL   http://ser-loopp.tc.cornell.edu/loopp.html
The LOOPP program (Learning, Observing and Outputting Protein Patterns) aligns sequence to sequence, sequence to structure (threading), and structure to structure. It can be used for the optimization of potentials and scoring functions of the above-mentioned applications.

PROSPECT

URL   http://compbio.ornl.gov/structure/prospect/
PROSPECT (PROtein Structure Prediction and Evaluation Computer Toolkit; Xu and Xu, 2000) is a threading prediction program that can be downloaded (free of charge for academic purposes) and installed

on Unix and Linux systems. The program performed well at the CASP4 competition (sixth out of 123 fold recognition competitors). PROSPECT automatically threads the target sequence (in FASTA or other formats) against a database of templates derived from the FSSP database. Depending on the hosting computer, this job may take 20 minutes or more than an hour. The program can be launched from command line, producing as output a sorted list of folds and information about the energy of the system.

PredictProtein server

URL   http://cubic.bioc.columbia.edu/predictprotein/
PredictProtein is an automatic tool for sequence analysis (multiple alignments, pattern matching, domains assignments) and structure prediction (it can infer secondary structure, transmembrane helices, solvent accessibility, coiled-coil regions and the fold adopted with a threading technique) of a target protein, using several different methods.

SAUSAGE

URL   http://rsc.anu.edu.au/~drsnag/TheSausageMachine.html
SAUSAGE (Sequence-structure Alignment Using a Statistical Approach Guided by Experiment, Huber *et al.*, 1999) is a threading program that can calculate protein sequence-structure alignments and search structure libraries. Different knowledge-based force fields are used for alignment calculations and subsequent ranking of calculated models. The approach adopted in SAUSAGE is to split the prediction calculation into separate steps of sequence-structure alignment and structure ranking.

10.4.6 *Ab initio* protein structure prediction

CORNET

URL   http://prion.biocomp.unibo.it/cornet.html

The CORNET procedure (Fariselli and Casadio, 1999) is a neural network based predictor that uses as input correlated mutations (Olmea and Valencia, 1997), sequence conservation, predicted secondary structure and evolutionary information. The system learns the association rules between the three-dimensional structure and its correspondent contact map, through the analysis of a set of known structures.

I-sites/Rosetta

URL   http://honduras.bio.rpi.edu/~isites/ISL_rosetta.html

The I-sites/Rosetta server performs *ab initio* predictions, expressed as backbone torsion angles, using a library of sequence-structure motifs. The procedure uses Rosetta, a Monte Carlo fragment insertion protein folding program (Simons *et al.*, 1997), and a hidden Markov model (HMMSTR, Bystroff *et al*, 2000) for local and secondary structure prediction, based on the I-sites library.

PDG-contact-pred

URL   http://montblanc.cnb.uam.es/cnb_pred/cgi-bin/
contact_pred_cgi

The PDG-contact-pred server predicts contacts between residues in the three-dimensional structure of a protein, using the correlated mutations approach. The crucial step is the construction of the multiple alignment. BLAST is used to search for homologous proteins in a non-redundant database, then the CLUSTALW multiple alignment program is used to align homologous sequences. The alignment is filtered to avoid redundancy, but very distant homologues are also eliminated.

10.4.7 Transmembrane protein prediction

DAS

URL   http://www.sbc.su.se/~miklos/DAS/
The DAS (Dense Alignment Surface; Cserzo *et al.*, 1997) method performs transmembrane helices prediction by comparing the query sequence with a non-redundant set of well-characterized membrane proteins. The prediction output has a textual and a graphical form, which can be downloaded in postscript format.

HMMTOP

URL   http://www.enzim.hu/hmmtop/index.html
HMMTOP (Hidden Markov Model for Topology Prediction) is a Hidden Markov Models-based method for transmembrane helices prediction. The procedure can be used on-line; moreover, the source code can be freely downloaded. Target sequences can be submitted in FASTA or NBRF/PIR formats, but unformatted sequences are also accepted.

MEMSAT

URL   http://www.biology.ucsd.edu/~yzhai/memsat.html
The MEMSAT (MEMbrane protein Structure And Topology) procedure is able to infer secondary structure and topology of membrane proteins composed of helix bundles. The prediction relies on data derived from characterized membrane proteins. The latest version of the algorithm, MEMSAT 2, is part of the PSIPRED server (URL http://insulin.brunel.ac.uk/psipred/).

MPEx

URL   http://blanco.biomol.uci.edu/mpex/index.html#Features Glance

The Membrane Protein Explorer (MPEx) server is able to analyze the topology and other features of membrane proteins, using experiment-based scales (the Wimley-White hydrophobicity scale; White and Wimley, 1999) to produce hydropathy plots. The user interface is a Java applet; which is also available as a stand-alone application that runs on MacOS, Windows and UNIX operating systems with a Java 1.1 or compatible Java Virtual Machine (JVM) / Java Runtime Environment (JRE).

TMHMM

URL   http://www.cbs.dtu.dk/services/TMHMM-2.0/

TMHMM is a Hidden Markov models-based procedure able to predict membrane protein topology and to discriminate between membrane and soluble proteins; it is thought to be one of the most reliable resources, being the one with the best overall performance in a systematic evaluation of the most widely used protein transmembrane region prediction methods (Möller *et al.*, 2001). The target sequence must be submitted in FASTA format; results are quickly available, and a sample graphical output is shown in Figure 10.9.

TopPred 2

URL   http://bioweb.pasteur.fr/seqanal/interfaces/toppred.html

TopPred is a membrane Protein Structure Prediction that relies on hydrophobicity analysis and the "positive inside" rule (Claros and von Heijne, 1994); the procedure allows the user to choose among different hydrophobicity scales and to vary some parameters, as the window length. The target sequence can be submitted in several formats.

10.4.8 Other useful links

EVA

URL   http://pipe.rockefeller.edu/~eva/
EVA (EValuation of Automatic protein structure prediction) project aim is to provide a continuous and fully automated analysis of structure prediction servers. Results are updated monthly, as new sequences are submitted to the servers (more or less twenty each week).

LiveBench

URL   http://bioinfo.pl/LiveBench/
The Live Bench Project is a continuous evaluation program of the structure prediction servers, especially for fold recognition resources.

Protein Structure Prediction Center

URL   http://predictioncenter.llnl.gov
The Center has organized the four Critical Assessment of techniques for protein Structure Prediction (CASP) competitions, and collects the results evaluated by experts in the field.

10.5 References

Alexandrov, N.N. 1996. SARFing the PDB. Protein Eng. 9: 727-732.

Anfinsen, C.B., Haber, E., Selas, M., and White, F.H. 1961. The kinetics of formation of native ribonuclease during oxidation of the reduced polypeptide chain. Proc. Natl. Acad. Sci. USA. 47: 1307-1314.

Anfinsen, C.B. 1973. Principles that govern the folding of protein chains. Science. 181: 233-239.

Bonneau, R. and Baker, D. 2001. Ab initio protein structure prediction: progress and prospects. Annu. Rev. Biophys. Biomol. Struct. 30: 173-189.

Brooks, B.R., Bruccoleri, R.E., Olafson, B.D., States, D.J., Swaminathan, S. and Karplus, M. 1983. CHARMM: A Program for Macromolecular Energy, Minimization, and Dynamics Calculations. J. Comp. Chem. 4: 187-2110.

Bryant, S.H. 1996. Evaluation of threading specificity and accuracy. Proteins. 26: 172-185.

Bystroff, C., Thorsson, V. and Baker, D. 2000. HMMSTR: a hidden Markov model for local sequence-structure correlations in proteins. J. Mol. Biol. 301: 173-190.

Bujnicki, J.M., Elofsson, A., Fischer, D. and Rychlewski, L. 2001 LiveBench-1: continuous benchmarking of protein structure prediction servers. Protein Sci. 10: 352-361.

Chandonia, J.M. and Karplus, M. 1999. New methods for accurate prediction of protein secondary structure. Proteins 35: 293-306.

Chothia, C. and Lesk, A.M. 1986. The relation between the divergence of sequence and structure in proteins. EMBO J. 5: 823-826.

Chothia, C. 1992. One thousand folds for the molecular biologist. Nature. 357: 543-544.

Chou, P.Y. and Fasman, G.D. 1974a. Prediction of protein conformation. Biochemistry. 13: 222-245.

Chou, P.Y. and Fasman, G.D. 1974b. Conformational parameters for amino acids in helical, beta-sheet, and random coil regions calculated from proteins. Biochemistry. 13: 211-222.

Chung, S.Y. and Subbiah, S. 1996. How similar must a template protein be for homology modeling by side-chain packing methods. In Proceedings of the first Pacific Symposium on Biocomputing. Hunter, L. and Klein, T., eds. Hawaii, USA p. 126-141.

Cohen, F.E., Sternberg, M.J. and Taylor, W.R. 1980. Analysis and prediction of protein beta-sheets structures by a combinatorial approach. Nature. 285: 378-382.

Cohen, F.E., Sternberg, M.J. and Taylor, W.R. 1982. Analysis and prediction of the packing of alpha-helices against a beta-sheet in the tertiary structure of globular proteins. J. Mol. Biol. 156: 821-862.

Claros, M.G. and von Heijne, G. 1994. TopPred II: an improved software for membrane protein structure predictions. Comput. Appl. Biosci. 10: 685-686.

Connolly, M.L. 1983. Solvent-accessible surfaces of proteins and nucleic acids. Science. 221: 709-713.

Cserzo, M., Wallin, E., Simon, I., von Heijne, G. and Elofsson, A. 1997. Prediction of transmembrane alpha-helices in prokaryotic membrane proteins: the dense alignment surface method. Protein Eng. 10: 673-676.

Cuff, J.A., Clamp, M.E., Siddiqui, A.S., Finlay, M. and Barton, G.J. 1998. JPred: a consensus secondary structure prediction server. Bioinformatics. 14: 892-893.

Cuff, J.A. and Barton, G.J. 2000. Application of multiple sequence alignment profiles to improve protein secondary structure prediction. Proteins. 40: 502-511.

Dayringer, H.E., Tramontano, A., Sprang, S.R. and Fletterick, R.J. 1986. Interactive program for visualization and modeling of proteins, nucleic acids and small molecules. J. Mol. Graphics. 4: 82-810.

Deane, C.M. and Blundell, T.L. 2001. CODA: a combined algorithm for predicting the structurally variable regions of protein models. Protein Sci. 10: 599-612.

Diederichs, K., Freigang, J., Umhau, S., Zeth, K. and Breed, J. 1998. Prediction by a neural network of outer membrane beta-strand protein topology. Protein Sci. 7: 2413-2420.

Di Francesco, V., Munson, P.J. and Garnier, J. 1999. FORESST: fold recognition from secondary structure predictions of proteins Bioinformatics. 15: 131-140.

Dunbrack, R.L. and Karplus, M. 1993. Backbone-dependent rotamer library for proteins. Application to side-chain prediction. J. Mol. Biol. 230: 543-574.

Edelman, J. 1993. Quadratic minimization of predictors for protein secondary structure. Application to transmembrane alpha-helices. J. Mol. Biol. 232: 165-191.

Eisenberg, D., Luthy, R. and Bowie, J.U. 1997. VERIFY3D: assessment of protein models with three-dimensional profiles. Meth. Enzymol. 277: 396-404.

Eisenhaber, F., Frommel, C. and Argos, P. 1996. Prediction of secondary structural content of proteins from their amino acid composition alone. I. New analytic vector decomposition methods. Proteins. 25: 157-168.

Engelman, D.M., Steitz, T.A. and Goldman, A. 1986. Identifying nonpolar transbilayer helices in amino acid sequences of membrane proteins. Annu. Rev. Biophys. Biophys. Chem. 15:321-353.

Finkelstein, A.V. 1997. Protein structure: what is possible to predict now? Curr. Op. Struct. Biol. 7: 60-71.

Fariselli, P. and Casadio, R. 1999. A neural network based predictor of residue contacts in proteins. Protein Eng. 12: 15-21.

Fischer, D., Barret, C., Bryson, K., Elofsson, A., Godzik, A., Jones, D., Karplus, K.J., Kelley, L.A., MacCallum, R.M., Pawowski, K., Rost, B., Rychlewski, L. and Sternberg, M. 1999. CAFASP-1: Critical assessment of fully automated structure prediction methods. Proteins 37 (S3): 209-2110.

Fischer, D. 2000. Hybrid fold recognition: combining sequence derived properties with evolutionary information. Pacific Symp. Biocomputing, Hawaii, 119-130.

Flores, T.P., Orengo, C.A., Moss, D.S. and Thornton, J. M. 1993. Comparison of conformational characteristics is structurally similar protein pairs. Protein Sci. 3: 2358-2365.

Frishman, D. and Argos, P. 1997. Seventy-five percent accuracy in protein secondary structure prediction. Proteins. 27: 329-335.

Frishman, D and Argos, P. 1996. Incorporation of non-local interactions in protein secondary structure prediction from the amino acid sequence. Protein Eng. 9: 133-142.

Garnier, J., Osguthorpe, D.J. and Robson, B. 1978. Analysis of the accuracy and implications of simple methods for predicting the secondary structure of globular proteins. J. Mol. Biol. 120:97-120.

Gibrat, J.F., Garnier, J. and Robson, B. 1987. Further developments of protein secondary structure prediction using information theory. New parameters and consideration of residue pairs. J. Mol. Biol. 198: 425-443.

Gobel, U., Sander, C., Schneider, R. and Valencia, A. 1994. Correlated mutations and residue contacts in proteins. Proteins. 18: 309-3110.

Goffeau, A., Slonimski, P., Nakai, K. and Risler, J.L. 1993. How many yeast genes code for membrane-spanning proteins? Yeast. 9: 691-702.

Gromiha, M.M., Majumdar, R. and Ponnuswamy, P.K. 1997. Identification of membrane spanning beta strands in bacterial porins. Protein Eng. 10: 497-500.

Guex, N. and Peitsch, M.C. 1997. SWISS-MODEL and the Swiss-

PdbViewer: An environment for comparative protein modeling. Electrophoresis. 18: 2714-2723.

Hilbert, M., Bohm, G. and Jaenicke, R. 1993. Structural relationships of homologous proteins as a fundamental principle in homology modeling. Proteins. 17: 138-151.

Hinds, D.A. and Levitt, M. 1994. Exploring conformational space with a sample lattice model for protein structure. J. Mol. Biol. 243: 668-682.

Holm, L. and Sander, C. 1993. Protein structure comparison by alignment of distance matrices. J. Mol. Biol. 233: 123-138.

Holm, L. and Sander, C. 1996. Mapping the protein universe. Science 273: 595-602.

Holm, L. and Sander, C. 1997. Dali/FSSP classification of three-dimensional protein folds. Nucl. Acids Res. 25: 231-234.

Holm, L. and Sander, C. 1998. Dictionary of recurrent domains in protein structures. Proteins. 33: 88-96.

Holm, L. and Sander, C. 1999. Protein fold and families: sequence and structure alignments. Nucl. Acid Res. 27: 244-2410.

Huber, H., Russell, A.J., Ayers, D. and Torda, A.E. 1999. SAUSAGE: Protein threading with flexible force fields. Bioinformatics. 15: 1064-1065.

Hutchinson, E.G. and Thornton, J.M.1996. PROMOTIF-a program to identify and analyze structural motifs in proteins. Protein Sci. 5: 212-220.

Jayasinghe, S., Hristova, K. and White, S.H. 2001. MPtopo: A database of membrane protein topology. Protein Sci. 10: 455-458.

Jeanteur, D., Lakey, J.H. and Pattus, F. 1991. The bacterial porin superfamily: sequence alignment and structure prediction. Mol. Microbiol. 5: 2153-2164.

Jones, D.T., Taylor, W.R. and Thornton, J.M. 1994. A model recognition approach to the prediction of all-helical membrane protein structure and topology. Biochemistry. 33: 3038-3049.

Jones, D.T. and Thornton, J. 1995. Potential energy functions for threading. Curr. Op. Struct. Biol. 6: 210-216.

Jones, D.T. 1997. Progress in protein structure prediction. Curr. Op. Struct. Biol. 7: 377-3810.

Jones, D.T. 1999. Protein secondary structure prediction based on position-specific scoring matrices. J. Mol. Biol. 292 :195-202.

Jones, D.T. 1999. GenTHREADER: an efficient and reliable protein fold recognition method for genomic sequences. J. Mol. Biol. 287: 797-815.

Jones, D.T., Tress, M., Bryson, K. and Hadley C. 1999. Successful recognition of protein folds using threading methods biased by sequence similarity and predicted secondary structure. Proteins. 37: 104-111.

Kabsch, W. and Sander, C. 1983. Dictionary of Protein Secondary Structure: pattern recognition of hydrogen-bonded and geometrical features. Biopolymers. 22: 2577-26310.

Kelley, L.A., MacCallum, R.M. and Sternberg, M.J. 2000. Enhanced genome annotation using structural profiles in the program 3D-PSSM. J. Mol. Biol. 299: 499-520.

Koretke, K.K., Russell R.B., Copley R.R. and Lupas, A.N. 1999. Fold recognition using sequence and secondary structure information. Proteins. 37: 141-148.

Krogh, A., Brown, M., Mian, I..S, Sjolander, K. and Haussler, D. 1994. Hidden Markov models in computational biology. Applications to protein modeling. J. Mol. Biol. 235: 1501-1531.

Krogh, A., Larsson, B., von Heijne, G. and Sonnhammer, E.L. 2001. Predicting transmembrane protein topology with a hidden Markov model: application to complete genomes. J. Mol. Biol. 305: 567-580.

Kyte, J. and Doolittle, R.F. 1982. A simple method for displaying the hydropathic character of a protein. J. Mol. Biol. 157: 105-132.

Laskowski, R.A., MacArthur, M.W., Moss, D.S. and Thornton, J.M. 1993. PROCHECK: a program to check the stereochemical quality of protein structures. J. Appl. Crys. 26: 283-291.

Laskowski, R.A. 2001. PDBsum: summaries and analyses of PDB structures. Nucl. Acids Res. 29: 221-222.

Lemer, C.M., Rooman, M.J. and Wodak, S.J. 1995. Protein structure prediction by threading methods: evaluatio of current techniques. Proteins. 23: 337-355.

Lim, V.I. 1974. Structural principles of the globular organization of protein chains. A stereochemical theory of globular protein secondary structure. J. Mol. Biol. 88: 857-872.

Lo Conte, L., Ailey, B., Hubbard, T.J., Brenner, S.E., Murzin, A.G. and Chothia, C. 2000. SCOP: a structural classification of proteins database. Nucl. Acids Res. 28: 257-259.

Lu, G. 2000. TOP: A new method for protein structure comparisons and similarity searches. J. Appl. Cryst. 33 : 176-183.

Lund, O., Frimand, K., Gorodkin, J., Bohr, H., Bohr, J., Hansen, J. and Brunak, S. 1997. Protein distance constraints predicted by neural networks and probability density functions. Protein Eng. 10: 1241-1248.

Lupas, A. 1996. Prediction and analysis of coiled-coil structures. Meth. Enzymol. 266: 513-525.

Luthy, R., Bowie, J.U. and Eisenberg, D. 1992. Assessment of protein models with three-dimensional profiles. Nature. 356: 83-85.

McGregor, M.J., Flores, T.P. and Sternberg, M.J.E. 1989. Prediction of beta-turns in proteins using neural networks. Protein Eng. 2: 521-526.

McGuffin, L.J., Bryson, K. and Jones, D.T. 2000. The PSIPRED protein structure prediction server. Bioinformatics. 16: 404-405.

Miyazava, S. and Jernigan, R.L. 1996. Residue-residue potentials with a favourale contact pair term and an unfavourable high packing density term, for simulation and threading. J. Mol. Biol. 256: 623-644.

Mizuguchi, K., Deane, C.M., Blundell, T.L. and Overington, J.P. 1998. HOMSTRAD: a database of protein structure alignments for homologous families. Protein Sci. 7: 2469-2471.

Möller, S., Croning, M.D. and Apweiler, R. 2001. Evaluation of methods for the prediction of membrane spanning regions. Bioinformatics. 17: 646-653.

Moulth, J. 1996. The current state of the art in protein structure prediction. Curr. Opin. Biotechnol. 7: 422-4210.

Moulth, J. 1999. Predicting protein three-dimensional structure. Curr. Op. Biotechnol. 10: 583-588.

Nakashima, H., Nishikawa, K. and Ooi, T. 1986. The folding type of a protein is relevant to the amino acid composition. J. Biochem. 99:153-162.

Nicholls, A., Sharp, K.A. and Honig, B. 1991. Protein folding and association: insights from the interfacial and thermodynamic properties of hydrocarbons. Proteins. 11: 281-296.

Ogata, K. and Umeyama, H. 2000. An automatic homology modeling method consisting of database searches and simulated annealing.

J. Mol. Graph. Model. 18: 258-272.

Olmea, O. and Valencia, A. 1997. Improving contact predictions by the combination of correlated mutations and other sources of sequence information. Fold. Des. 2: 25-32.

Orengo, C.A., Brown, N.P. and Taylor, W.R. 1992. Fast structure alignment for protein databank searching. Proteins. 14: 139-1610.

Orengo, C.A., Jones, D.T. and Thornton, J.M. 1994. Protein superfamilies and domain superfolds. Nature. 372: 631-634.

Orengo, C.A., Michie, A.D., Jones, S., Jones, D.T., Swindells, M.B., and Thornton, J.M. 1997. CATH- A Hierarchic Classification of Protein Domain Structures. Structure. 5: 1093-1108.

Ortiz, A.R., Kolinski, A. and Skolnick, J. 1998. Fold assembly of small proteins using Monte Carlo simulations driven by restraints derived from multiple sequence alignments. J. Mol. Biol. 259: 349-365.

Park, B.H. and Levitt, M. 1995. The complexity and accuracy of discrete state models of protein structure. J. Mol. Biol. 249: 493-5010.

Pedersen, J.T. and Moulth, J. 1996. Genetic algorithms for protein structure prediction. Curr. Op. Struct. Biol. 6: 227-231.

Peitsch, M.C. 1996. ProMod and Swiss-Model: Internet-based tools for automated comparative protein modeling. Biochem. Soc. Trans. 24: 274-279.

Ponder, J.W. and Richards, F.M. 1987. Tertiary templates for proteins. Use of packing criteria in the enumeration of allowed sequences for different structural classes. J. Mol. Biol. 193: 775-791.

Qian, N. and Sejnowski, T.J. 1988. Predicting the secondary structure of globular proteins using neural network models. J. Mol. Biol. 202: 865-884.

Rabow, A.A. and Scheraga, H.A. 1996. Improved genetic algorithm for the protein folding problem by use of a Cartesian combination. Protein Sci. 5: 1800-1815.

Reva, B.A. and Finkelstein, A.V. 1996. Search for the most stable folds of protein chains: II. Computation of stable architectures of beta-proteins using a self-consistent molecular field theory. Protein Eng. 9: 399-411.

Rychlewski, L., Jaroszewski, L., Li, W. and Godzik, A. 2000. Comparison of sequence profiles. Strategies for structural predictions using sequence information. Protein Sci. 9: 232-241.

Rost, B. and Sander, C. 1993. Prediction of protein secondary structure at better than 70% accuracy. J. Mol. Biol. 232: 584-599.

Rost, B. 1996. PHD: predicting one-dimensional protein structure by profile-based neural networks. Meth. Enzymol. 266: 525-539.

Rost, B., Fariselli, P. and Casadio, R. 1996. Topology prediction for helical transmembrane proteins at 86% accuracy. Protein Sci. 5: 1704-1718.

Russell, R.B., Copley, R.R. and Barton, G.J. 1996. Protein fold recognition by mapping predicted secondary structures. J. Mol. Biol. 259: 349-365.

Russell, R.B., Saqi, M.A.S., Bates, P.A., Sayle, R.A. and Sternberg, M.J.E. 1998. Recognition of analogous and homologous folds - Assessment of prediction success and associated alignment accuracy using empirical matrices. Prot. Eng. 11: 1-9.

Salamov, A.A. and Solovyev, V.V. 1995. Prediction of protein secondary structure by combining nearest-neighbor algorithms and multiple sequence alignments. J. Mol. Biol. 247: 11-15.

Sali, A. and Blundell, T.L. 1993. Comparative protein modeling by satisfaction of spatial restraints. J. Mol. Biol. 234: 779-815.

Sanchez, R., Pieper, U., Mirkovic, N., deBakker, P.I.W., Wittenstein, E. and Sali, A. 2000. MODBASE, a database of annotated comparative protein structure models. Nucl. Acids Res. 28: 250-253.

Sander, C. and Schneider, R. 1991. Database of homology-derived protein structures and the structural meaning of sequence alignment. Proteins. 9: 56-68.

Sayle, R.A. and Milner-White, E.J. 1995. RASMOL: Biomolecular graphics for all. Trends Biochem. Sci. 20: 374-376.

Schirmer, T. and Cowan, S.W. 1993. Prediction of membrane-spanning beta-strands and its application to maltoporin. Protein Sci. 2: 1361-1363.

Shi, J., Blundell, T.L. and Mizuguchi K. 2001. FUGUE: sequence-structure homology recognition using environment-specific substitution tables and structure-dependent gap penalties. J. Mol. Biol. 310: 243-2510.

Simons, K.T., Kooperberg, C., Huang, E. And Baker, D. 1997. Assembly of proein tertiary structures from fragments with similar local sequences using simulated annealing and Bayesian scoring functions. J. Mol. Biol. 268: 209-225.

Simons, K.T., Bonneau, R., Ruczinski, I. and Baker, D. 1999. Ab initio protein structure prediction of CASP III targets using ROSETTA. Proteins. 37: 171-176.

Skolnick, J. and Kolinski, A. 1991. Dynamic Monte Carlo simulations of a new lattice model of globular protein folding, structure and dynamics. J. Mol. Biol., 221:499-531.

Sternberg, M.J., Bates, P.A., Kelley, L.A. and MacCallum, R.M 1999. Progress in protein structure prediction: assessment of CASP3. Curr. Opin. Struct. Biol. 9: 368-373.

Stevens, T.J. and Arkin, I.T. 2000. Do more complex organisms have a greater proportion of membrane proteins in their genomes? Proteins. 39: 417-420.

Taylor, W.R. and Orengo, C.A. 1989. Protein structure alignment. J. Mol. Biol. 208: 1-22.

Taylor, W.R. 1999. Protein structure comparison using iterated double dynamic programming. Protein Sci. 8: 654-665.

Thomas, D.J., Casari, G. and Sander, C. 1997. The prediction of protein contacts from multiple sequence alignments. Protein Eng. 9: 941-948.

Tusnady, G.E. and Simon, I. 1998. Principles governing amino acid composition of integral membrane proteins: application to topology prediction. J. Mol. Biol. 283: 489-506.

van Gunsteren W.F., Brunne R.M., Gros P., van Schaik R.C., Schiffer C.A. and Torda A.E. 1994. Accounting for molecular mobility in structure determination based on nuclear magnetic resonance spectroscopic and X-ray diffraction data. Meth. Enzymol. 239: 619-654.

von Heijne, G. and Gavel, Y. 1988. Topogenic signals in integral membrane proteins. Eur. J. Biochem. 174: 671-678.

Vriend, G. and Sander, C. 1991. Detection of common three-dimensional substructures in proteins. Proteins. 11: 52-58.

Wallin, E. and von Heijne, G. 1998. Genome-wide analysis of integral membrane proteins from eubacterial, archaean, and eukaryotic organisms. Protein Sci. 7: 1029-1038.

Walther, D. 1997. WebMol - a Java-based PDB viewer. Trends Biochem. Sci. 22: 274-275.

Wallace, A.C., Laskowski, R.A. and Thornton, J.M. 1995. LIGPLOT:

A program to generate schematic diagrams of protein-ligand interactions. Prot. Eng. 8: 127-134.

Weiner, P.K. and Kollman, P.A. 1981. AMBER: Assisted Model Building with Energy Refinement. A General Program for Modeling Molecules and Their Interactions. J. Comp. Chem. 2: 2810.

Welte, W., Weiss, M.S., Nestel, U., Weckesser, J., Schiltz, E. and Schulz, G.E. 1991. Prediction of the general structure of OmpF and PhoE from the sequence and structure of porin from Rhodobacter capsulatus. Orientation of porin in the membrane. Biochim. Biophys. Acta 1080: 271-274.

Wilmanns, M. and Eisenberg, D. 1995. Inverse protein folding by the residue pair preference method: estimating the correctness of alignments of structurally compatible sequences. Protein Eng. 8: 627-639.

Wimley, W.C. and White, S.H. 1996. Experimentally determined hydrophobicity scale for proteins at membrane interfaces. Nat. Struct. Biol. 3: 842-848.

Wolf, E., Kim, P.S. and Berger, B. 1997. MultiCoil: a program for predicting two- and three-stranded coiled coils. Protein Sci. 6: 1179-1189.

Xu, Y. and Xu, D. 2000. Protein threading using PROSPECT: design and evaluation. Proteins. 40 : 343-354.

Zvelebil, M.J., Barton, G.J., Taylor, W.R. and Sternberg, M.J. 1987. Prediction of protein secondary structure and active sites using the alignment of homologous sequences. J. Mol. Biol. 195: 957-961.

11

Let Others Solve your Problems: the Newsgroups

Richard P. Grant

Contents

Abstract

Newsgroups permit individuals to take part in a worldwide discussion on a specific topic of interest. A message is "posted" to a newsgroup usually by email or web form. Any other member of that discussion group can read and reply to the message. The BIOSCI bionet newsgroup network allows easy communication between life scientists world wide. This chapter provides a complete listing and a brief description of the bionet newsgroups and describes in detail the use of these newsgroups via a web browser and through dedicated news reader software.

From: *The Internet for Cell and Molecular Biologists: Current Applications and Future Potential*
ISBN 1-898486-32-8 © 2002 Horizon Scientific Press, Wymondham, UK

11.1 Usenet for beginners

Although they do not have the 'glamour' of the World Wide Web (WWW), and many users are unaware of their existence, newsgroups constitute a major proportion of traffic carried over the internet. These newsgroups make up 'Usenet', a world-wide distributed discussion system. The newsgroups themselves have names that are classified hierarchically by subject. They are similar to email in that a person with a computer, the necessary software and an internet connection can contribute to them and receive replies; they are dissimilar in that an email is sent to one or a few known people whereas a message (or article) posted to a Usenet newsgroup can, in theory, be read by anyone in the world.

Usenet had its beginnings in the late 1970s, when graduate students at the University of North Carolina thought of hooking computers together to exchange information with the Unix community. Over the next few years more computers (or 'sites') were added to the nascent network until enough were organized and linked together to give most of North America access. Usenet as we know it today really came into being around 1987, when the alt.*, comp.*, misc.*, news.*, rec.*, sci.*, soc.*, talk.* and local hierarchies were formed, and reliable links to Europe were established. More networks and hierarchies, including bionet.*, were added later.

Usenet performs a completely different function to the WWW. A web site is essentially a means for the author (be they an individual, a company, or an organization) to pass information to the rest of the world. Unlike the WWW, information on Usenet flows both (or many) ways. A web site is like a plenary lecture at a conference. A newsgroup on the other hand is more analogous to a round table discussion, or perhaps chatting in the conference centre bar. There may be a discussion leader, or there may not. The people involved may all be friends, or colleagues, or complete strangers. Participants may even have to make their comments through a third party for approval.

In terms of propagation, again Usenet is different from the WWW. Web pages are stored on a single server, requests are made from the user's computer for pages, graphics, files, etc. which are then sent to the user. The transmitted information may travel over international networks but any given web site has only one physical location. The posted messages or articles that make up a newsgroup however are transmitted from the originator to and stored on many thousands of news servers around the world. Therefore not all newsgroups are necessarily available to everyone - the administrators of news servers are able to exercise a great deal of control over which newsgroups are available to their users. The mainstream hierarchies are generally carried and distributed without interference, and newsgroups of local or national interest are usually not carried by news servers outside of the appropriate locality or country.

It is worth pointing out that the content (i.e. individual articles) of newsgroups is not censored by news administrators, but rather the decision to carry or not is made at the level of the newsgroup itself. When a group is not carried by a server it is not necessarily that there is an objection to its content, but more prosaically that the users at that site just are not interested in it. Some newsgroups are 'moderated'; this generally takes the form that articles are sent first to a 'moderator' for approval and/or editing before appearing in the newsgroup. The moderator may be one or more people, and has an active interest in the newsgroup that is being moderated. It is incorrect to classify this as 'censorship'; for example some newsgroups are for announcements only, others may have a very technical nature and in such cases it is appropriate to disallow general chat or 'off-topic' postings. Moderated newsgroups tend to have a better signal-to-noise ratio than those that are not moderated. Importantly, decisions which affect the Usenet community - such as whether to create or moderate a group - are made and implemented by the community itself. Thus a user at a site which does not carry a particular newsgroup may well be able to persuade the administrator to add it to his server.

11.2 Bionet

The Bionet.* news hierarchy contains the BIOSCI newsgroups. The BIOSCI network allows easy communication between life scientists world wide. As with the rest of Usenet, this is independent of the computer systems or networks used. To provide access to the newsgroups for those without access to a news server, BIOSCI operates a mail to news gateway so that articles posted to a bionet newsgroup are relayed to an equivalent mailing list, and vice versa. All messages are archived on the WWW at http://www.bio.net/, which also provides a searchable interface to all the groups. Although the number of scientists who actively use Bionet is probably quite small compared with the total number of practising life scientists world wide, there is a wealth of knowledge and experience available. Even if the active contributors do not know the answer to a query, the chances are that they know where to look or who to ask.

BIOSCI is supported by advertising sponsorships at <http://www.bio.net/> in the USA, and by the Medical Research Council's HGMP-RC in the UK. Thus access to Bionet is free of charge. Although the groups are available to all, they are for communication between researchers, and are not intended to provide a forum or a consulting service for lay persons. The sci.* hierarchy is appropriate in such cases.

Some of the more specialized Bionet newsgroups, including those for the professional societies, are moderated. Announcements of interest to the entire BIOSCI community are made in bionet.announce, which is also moderated. The open discussion forum is bionet.general, but probably the group which people find most useful is bionet.molbio.methds-reagnts. This is a forum for exchanging protocols and requests for reagents and (methodological) help. It is unmoderated and has a relatively high level of traffic.

Table 11.1 provides a complete listing of the Bionet newsgroups and a brief description (correct as of November 2001). Moderated groups are indicated. Equivalent mailing list names and names of the discussion leaders and moderators may be obtained from the BIOSCI electronic newsgroup network information sheet at <http://www.bio.net/docs.html>.

11.3 Access and (n)etiquette

It is possible to use a web browser to read and post news, but it is preferable to use a dedicated news reader. Such dedicated software is more likely to have a useful feature set for Usenet reading and posting, allow off-line or on-line[1] browsing (depending on the reader) and make

---

[1]on-line: a connection to the internet is maintained while reading and posting. Recommended for those with a permanent internet connection, or unmetered modem access.

Table 11.1. A complete listing and a brief description of the Bionet newsgroups.

bionet.agroforestry	Discussions about agroforestry research.
bionet.announce	Announcements of widespread interest to life scientists. (Moderated)
bionet.audiology	Topics in audiology and hearing science. (Moderated)
bionet.biology.cardiovascular	Research discussions in cardiovascular biology.
bionet.biology.computational	Mathematical and computer applications in biology. (Moderated)
bionct.biology.deepsea	Deep-sea marine biology, oceanography and geology research. (Moderated)
bionet.biology.grasses	Research into the biology of grasses, especially cereal, forage, and turf species.
bionet.biology.n2-fixation	Biological nitrogen fixation research.
bionet.biology.symbiosis	Research in symbiosis (Moderated)
bionet.biology.tropical	Research in tropical biology.
bionet.biology.vectors	Research and control of arthropods which transmit disease. (Moderated)
bionet.biophysics	The science and profession of biophysics.
bionet.celegans	Research discussions on *Caenorhabditis elegans* and related nematodes. (Moderated)
bionet.cellbiol	Discussions about cell biology including cancer research at the cellular level.
bionet.cellbiol.cytonet	Discussions about research on the cytoskeleton, plasma membrane and cell wall.

bionet.cellbiol.insulin	Discussions about the biology and chemistry of insulin and related receptors. (Moderated)
bionet.chlamydomonas	Research discussions about the biology of the green alga *Chlamydomonas* and related genera. (Moderated)
bionet.diagnostics	Problems and techniques in all fields of diagnostics. (Moderated)
bionet.diagnostics.prenatal	Discussions about research in prenatal diagnostics.
bionet.drosophila	Research into the biology of fruit flies. (Moderated)
bionet.ecology.physiology	Research and education in physiological ecology. (Moderated)
bionet.emf-bio	Discussions about research on electromagnetic field interactions with biological systems. (Moderated)
bionet.general	Discussions about biological topics for which there is not yet a dedicated newsgroup, and general BIOSCI discussion.
bionet.genome.arabidopsis	Information about the *Arabidopsis* genome project. (Moderated)
bionet.genome.autosequencing	Research and support on automated DNA sequencing. (Moderated)
bionet.genome.chromosomes	Discussions about mapping and sequencing of eukaryote chromosomes.
bionet.genome.gene-structure	Genome and chromatin structure and function. (Moderated)
bionet.genomes.markers	Molecular markers, microsatellites and AFLPs. (Moderated)
bionet.glycosci	Discussions about carbohydrate and glycoconjugate molecules.
bionet.immunology	Research in immunology.

bionet.info-theory	'Biological Information - Theory And Chowder Society': Applications of Information theory to biology; this group is not for general information or for discussion of theories in general.
bionet.jobs.offered	Job openings in the biological sciences. (Moderated)
bionet.jobs.wanted	Forum for posting resumes/CVs by individuals seeking employment in the biological sciences or in support of the biological sciences.
bionet.journals.letters.biotechniques	Discussions of articles in the journal Biotechniques. (Moderated)
bionet.journals.note	Practical advice on dealing with professional biological journals.
bionet.maize	Research on maize. (Moderated)
bionet.metabolic-reg	Discussions about the kinetics, thermodynamics and control of biological processes at the cellular level.
bionet.microbiology	The science and profession of microbiology.
bionet.microbiology.biofilms	Research on microbial biofilms. (Moderated)
bionet.molbio.ageing	Research into cellular and organismal ageing.
bionet.molbio.bio-matrix	Application of computers to biological databases.
bionet.molbio.embldatabank	Messages to and from the EMBL nucleic acid database staff. (Moderated)
bionet.molbio.evolution	Discussions about research in molecular evolution. (Moderated)
bionet.molbio.gdb	Messages to and from the Genome Data Bank staff.
bionet.molbio.genbank	Messages to and from the GenBank

	nucleic acid database staff . (Moderated)
bionet.molbio.gene-linkage	Research into genetic linkage analysis.
bionet.molbio.genearrays	Gene array and micro array technology.
bionet.molbio.genome-program	NIH-sponsored newsgroup on human genome issues. (Moderated)
bionet.molbio.hiv	Research and discussions into the molecular biology of HIV.
bionet.molbio.methds-reagnts	Requests for information and lab reagents. The heart of bionet.
bionet.molbio.molluscs	Research on mollusc DNA. (Moderated)
bionet.molbio.proteins	Discussions about research on proteins and messages for the PIR and SWISS-PROT databank staffs.
bionet.molbio.proteins.7tms_r	Discussions about signal transducing receptors which interact with G-proteins. (Moderated)
bionet.molbio.proteins.fluorescent	Research on fluorescent proteins and bioluminescence.
bionet.molbio.rapd	Discussions about Randomly Amplified Polymorphic DNA.
bionet.molbio.recombination	Research on the recombination of DNA or RNA. (Moderated)
bionet.molbio.yeast	The molecular biology and genetics of Yeast. (Moderated)
bionet.molec-model	Physical and chemical aspects of molecular modelling.
bionet.molecules.free-radicals	Research on free radicals in biology and medicine. (Moderated)
bionet.molecules.p450	Research on cytochrome P450.
bionet.molecules.peptides	Research involving peptides. (Moderated)
bionet.molecules.repertoires	Generation and use of libraries of molecules. (Moderated)

bionet.mycology	Discussions about research on filamentous fungi. (Moderated)
bionet.neuroscience	Research issues in the neurosciences.
bionet.neuroscience.amyloid	Forum for researchers on Alzheimer's disease and related disorders, including prion diseases. (Moderated)
bionet.organisms.pseudomonas	Research on the genus *Pseudomonas*.
bionet.organisms.schistosoma	Discussions about *Schistosoma* research. (Moderated)
bionet.organisms.urodeles	Discussions about research in *Urodele* amphibian biology. (Moderated)
bionet.organisms.zebrafish	Discussions about research using the model organism Zebrafish (*Danio rerio*). (Moderated)
bionet.parasitology	Research into parasitology. (Moderated)
bionet.photosynthesis	Discussions about photosynthesis research. (Moderated)
bionet.plants	Research into plant biology.
bionet.plants.education	Discussion of education issues in plant biology. (Moderated)
bionet.plants.signaltransduc	Research on plant signal transduction. (Moderated)
bionet.population-bio	Population biology research.
bionet.prof-society.afcr	American Federation for Clinical Research. (Moderated)
bionet.prof-society.aibs	American Institute of Biological Sciences announcements. (Moderated)
bionet.prof-society.biophysics	Official announcements/information from the Biophysical Society. (Moderated)
bionet.prof-society.cfbs	Newsgroup for the Canadian

	Federation of Biological Societies (CFBS). (Moderated)
bionet.prof-society.csm	Announcements from the Canadian Society of Microbiologists. (Moderated)
bionet.prof-society.navbo	Forum for the North American Vascular Biology Organization. (Moderated)
bionet.protista	Discussions about ciliates and other protists (protozoa, algae, zoosporic fungi). (Moderated)
bionet.software	Information about software for the biological sciences. (Moderated)
bionet.software.acedb	Discussions by users and developers of genome databases using the ACEDB software. (Moderated)
bionet.software.gcg	Discussions about the GCG sequence Discussions about the Sequence Retrieval System (SRS) software. (Moderated)
bionet.software.staden	Discussions between scientists using the Staden molecular sequence analysis software.
bionet.software.www	Announcements about resources in biology which can be accessed via electronic networked information retrieval software. (Moderated)
bionet.software.x-plor	Discussions about the X-PLOR software for 3D macromolecular structure determination.
bionet.structural-nmr	Discussions about the use of NMR for macromolecular structure determination. (Moderated)
bionet.toxicology	Research in toxicology. (Moderated)
bionet.users.addresses	Who's who in Biology: Question/ answer forum for help using electronic networks, locating e-mail

	addresses, etc.
bionet.virology	Discussions about research in virology. (Moderated)
bionet.women-in-bio	Discussion of issues concerning women in the biological sciences. (Moderated)
bionet.xtallography	Discussion about crystallography of macromolecules and messages for the PDB staff.

it easier for the user to adhere to Usenet guidelines for posting (see <http://www.faqs.org/faqs/usenet/emily-postnews/part1/>). They are also more likely to have sophisticated filtering mechanisms and to be less susceptible to embedded scripts or viruses. Free and shareware software may be found at <http://www.newsreaders.com/>, or by searching for 'news reader' or 'newsreader' at <http://www.versiontracker.com/> (for Mac OS X and Classic), <http://download.cnet.com/> or <http://www.tucows.com/> (both for Mac, Linux and Windows). It is difficult to recommend any one program, as personal preferences pay a large part in software choice. However, on the Mac I use MT-NewsWatcher for on-line reading, but MacSoup is good if you want to read news off-line[2]. Forté Agent is a Windows-only commercial news reader, with a free version available. When on a Unix system, I have used *tin*, but *slrn* is also reported to be good.

The alternative to using dedicated news software is to use a web to news gateway. This is particularly useful for when you are away from your computer or want to read and post on a 'shared' machine. A popular gateway, with access to and archives of all newsgroups, is at <http://groups.google.com/>. Anyone can search and read messages, but to post your own messages you must register with a valid email address and provide a username and a password. This ensures that no

---

[2]off-line: the news reader connects to the internet, downloads the contents of the groups to which you are subscribed, then disconnects. When you have finished writing posts the reader re-connects to send them. This is good for those with a slow modem link or who want to limit their time on-line.

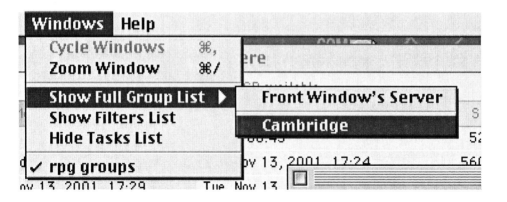

Figure 11.1. Choosing newsgroups with MT-NW.

one else can use your identity to post within a group. Another one is at <http://www.newsranger.com/>, but registration (again, this does not cost anything) is required for reading as well as posting. An extensive list of web-based news readers may be found at <http://directory.google.com/Top/Computers/Usenet/Web_Based/>. It is worth remembering that all web-based readers require an internet connection that is maintained for the duration of the session.

BIOSCI can be an incredibly useful resource, but its utility depends on users observing certain guidelines. After installing a news reader (and having read the accompanying documentation) it is tempting to leap straight into a group and post a question, or a test message. However, the newsgroups (both in Bionet.* and the rest of Usenet) have acquired a strong sense of society, with rules for good behaviour and a set of 'regulars'. In the real world it is considered polite to listen before speaking - especially when encountering a group of people that know each other - and similar etiquette is expected on Usenet. It is important to remember that there are real people behind the articles, and failure to observe custom can upset other users and make life generally unpleasant. The rules and customs of Usenet in general also apply to Bionet.*, particularly because the users are professionals and expect a professional standard of behaviour from everyone else. While people who regularly make use of the Bionet groups tend to be friendly and welcoming towards newcomers, asking for answers to a term paper (for example) can provoke unwanted

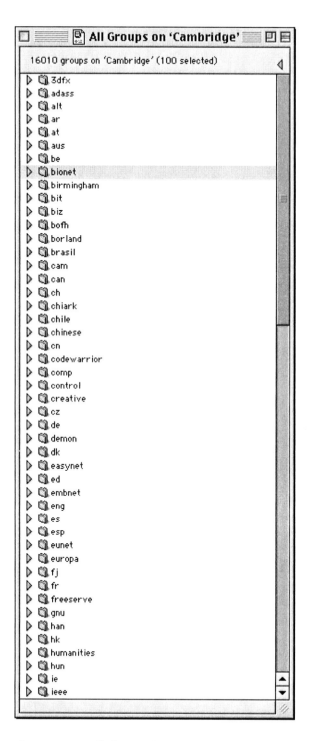

Figure 11.2. The entire newsgroup listing on the Cambridge University server.

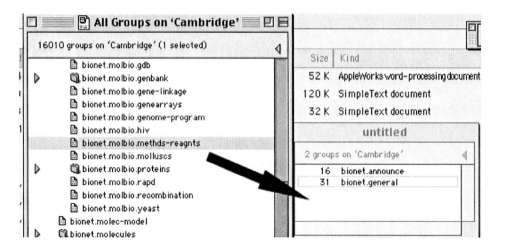

Figure 11.3. Dragging groups from the entire listing into a new group window.

reaction. Commercial advertising is also unwelcome, and breaks the written charter of BIOSCI. It also pays to check the searchable BIOSCI archive at <http://www.bio.net/> before making a post, just in case the question one wants to ask may have been answered only recently.

The above might sound intimidating, but reading the news.announce.newusers group and just browsing the groups that are of interest for a week or two before making any posts will enable most people to get a feel of what is expected. The small amount of effort required is small compared with the potential benefits. That said, let us now take a brief look at the mechanics of reading and posting news.

11.4 How to use a news reader

As mentioned above, there are multiple ways to access Usenet. It would be impossible to cover all news readers in detail, so what follows is a brief description of how to use MT-NewsWatcher (MT-NW: latest version available from <http://www.smfr.org/mtnw/>), a popular freeware reader for the Mac. The principles of use will be the same for most other graphical readers. When you launch the program for the first time you will be walked through a set-up process, during which you will fill in personal details, an email address that you don't

#	S	Authors	Subjects	Lines	Date & Tim
		Itabajara da Silva Vaz J...	Materilas & Methods Online	18	12:25
		Dr. Klaus Eimert	Chemiluminescence Imagers?	33	11:23
		me@there.com	req: any etexts	3	06:45
		Igor Evsikov	Scientific and engineering expresson calculator+grapher ∗ unit converter-/(MATRIX^COM...	59	Yesterday 22:06
△	4	teamclub1012@Flashmai...	GUARANTEED MONTHLY INCOME - Join FREE NOW!!	61	Yesterday 21:49
		Trond Erik Vee Aune	Help with quantitative analyses of protein	21	Yesterday 14:32
△	3	Silas Bruun	protocol web page down??	21	Yesterday 14:12
△		Kelvin	High CG sequencing	11	Yesterday 13:23
		Fun Science	Total RNA / DNA / Protein isolation reagents or kits	13	Yesterday 02:54
△		Biddle Consulting Group, ...	Validated Office Skills Testing	44	Yesterday 01:17
△	2	Stefan Roepcke	Composition and amount of mRNA's	19	Tuesday 21:12
△	6	Martin Canizales	Stable transformants using PCR amplified linear DNA	12	Tuesday 17:00
△	4	dbell	RE: Digestion problems with EcoRI	46	Tuesday 15:49
△	2	Martina and Klaus Lehnert	His-tagging positively charged proteins ???	30	Tuesday 06:43
		Ali Karami	Plasmid Copy number	46	Tuesday 06:24
△	12	Dr. James J. Campanella	Faculty position-Montclair State University, Bioinformatics	37	Monday 18:46
△	2	Frederik Wirtz-Peitz	PfuTurbo (hotstart version)	35	Monday 17:49
△		Itabajara da Silva Vaz J...	removal of nucleic acids.	17	Monday 12:59
		Barbara Simionati	Digestion problems with EcoRI	22	Monday 12:02
▷		takedatu	web site for protein	5	Monday 07:53
△	3	Sergio	Re: web site for protein	12	Monday 11:52

211 articles, 211 unread, 0 killed

Figure 11.4. Reading new messages from a newsgroup.

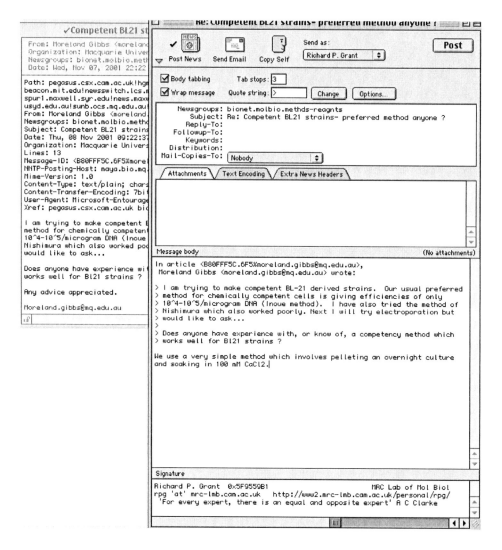

Figure 11.5. Posting a reply ('followup') to an article.

mind thousands of other people seeing, and a news server address.
This latter is the most difficult to get right. If you are at an academic
institute then the best bet is to find whence comes your institute's
news 'feed' and use that as your server. Search your local 'help' pages
or ask your systems administrator. As institutional servers usually
disallow users from outside their network then it is essential to get the
correct server address. For those not at an academic institution then
your internet service provider probably has a server that you can use -

check their web pages for help. A list of news servers that anyone can use can be found at <http://www.newzbot.com/>. At home I use the server at <http://news.fu-berlin.de/>, which is free but requires registration. Of course, using a web-based news reader this is all done for you.

Warning! This tutorial is no substitute for reading the manual!

Given that you have set up your news reader correctly (read the friendly manual for help) then the next step is to choose the newsgroups that interest you and 'subscribe' to them. With MT-NW this is done by choosing 'Show Full Group List' from the 'Windows' menu to bring up all the news groups on the server (Figure 11.1). With MT-NW it is possible to use multiple servers; in the example there is only one, which is called 'Cambridge'. This results in a window somewhat like Figure 11.2, which shows part of the entire newsgroup listing on the Cambridge University server, with the Bionet hierarchy selected. Clicking on the triangles expands each hierarchy. Next, one may open a new group window and simply drag groups from the entire listing into it (Figure 11.3; the numbers indicate how many articles there are in the groups). Once this is done, save the new group window and you have a document which contains groups to which you are subscribed.

To actually read news, select a group and open it - either by double clicking or hitting 'return'. You will see all the unread messages in the group (Figure 11.4). Messages can be read either by double-clicking on them or by using keyboard short-cuts for speed. 'Threads' or topics (multiple postings on the same subject) can be expanded by clicking on the triangles. Articles that you read are marked as read so that they do not appear again. Any articles you're not interested in can be 'marked as read' so that they do not appear. For other options, such as 'filtering' and 'killing', check the extensive help documentation.

Posting a reply to an article is known as posting a 'followup'. With MT-NW this function resides in the 'News' menu. When you make a followup (Figure 11.5), the original author's text appears in a window with quote symbols next to it, to indicate that you are quoting someone. Your own message should go beneath the quoted text. Trimming the quoted parts (as shown) is encouraged - this cuts down on unnecessary server load and saves readers of your articles having to wade through stuff they've already read. However, trimming too much can lead to confusion for people trying to follow a thread. Sufficient quoted text should remain so that it is clear that you are replying to someone else's post, and the attributions (the bit starting 'In article . . .') must be left in so that it is clear who said what. If you are replying to a long article with a number of issues it is standard practice to interleave your replies to each point with the original text, for clarity. Other things to note: your articles should be less than 80 characters wide, to allow for narrow screens and quoting - 72 characters is good. It is standard practice to use a fixed-width font, such as Courier or Monaco, as traditionally that is what most people will be using to read your posts. Related to this, for reasons of bandwidth and compatibility, all posts should be made in 'plain text'; i.e. **not** in HTML or formatted. A signature, which can be automatically included, should be no more than four lines long. Remember that your message potentially may be read by thousands of people worldwide, and write accordingly. That's it. Hit 'Post' and send your message (Figure 11.6).

11.5 Whither Bionet?

The Bionet hierarchy has mushroomed since its inception. Unfortunately its growth has brought with it a number of problems. It is very much a victim of its own success; some of the groups are so busy that it is impossible to keep abreast of them all (much like science in the real world). Cultivating a disciplined approach to Usenet is desirable: it is not necessary to read all the groups, and it certainly is not necessary to read all the posts in any given group.

A major difficulty is with 'spam'. This is the term for articles that are usually advertising a product or web site and posted to multiple news groups (see also Chapter 3). Spam constitutes a major proportion of Usenet traffic. Fortunately, much of this nonsense is filtered out at news server sites before it reaches the end user, and it is possible to make effective use of 'kill files' or the filtering capabilities of news reader software. With practise, a user will also be able to recognize the junk and delete it without reading. Perhaps the only way to regain the high signal to noise ratio of the early days is to moderate the Bionet hierarchy. This would involve a major commitment on the part of some users, and might adversely affect the speed and spontaneity of

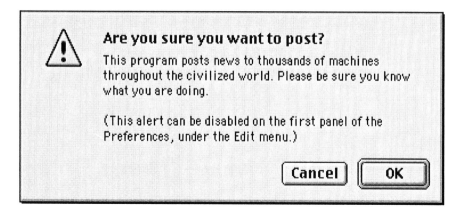

Figure 11.6. The final step to send your message.

the groups. It may be that the users decide that this is a price worth paying. Some groups in the Bionet.* hierarchy seem to have been completely taken over by spam, and scientists no longer appear to post to them. It may be time for the BIOSCI administrators to re-examine the entire concept of Bionet.

People have been predicting the 'death of the 'net' from the mid-80s. Despite this, Usenet thrives and bionet.molbio.methds-reagnts (for example) is by far the best resource available to researchers today for international exchange of laboratory advice, know-how, and ideas. Several years ago my doctorate supervisor told me that Bionet would not change my life. He was absolutely right. I have, however, found the answers to some of my own questions and been able to answer others'. Bionet facilitates, and long may it continue to do so.

11.6 Useful links and further reading

The news group news.announce.newusers is a must read for those new to Usenet. There is a frequently asked questions list (FAQ) at: <ftp://rtfm.mit.edu/pub/usenet/news.announce.newusers/ Answers_to_Frequently_Asked_Questions_about_Usenet>.

Acronyms and jargon are an integral part of Usenet, and some explanations may be found at: <http://www.tuxedo.org/~esr/jargon/> and <http://www.fun-with-words.com/acronyms.html>.

There is a canonical 'smiley' (or 'emoticon') list at:
<http://www.astro.umd.edu/~marshall/smileys.html>.

Hints on writing for Usenet can be found at
<ftp://rtfm.mit.edu/pub/usenet/news.announce.newusers/
Hints_on_writing_style_for_Usenet>. A guide to quoting style may
be found at <http://www.greenend.org.uk/rjk/2000/06/14/
quoting.html>.

The FAQ for BIOSCI/Bionet is at:
<http://www.bio.net/docs/biosci.FAQ.html>.

A FAQ for bionet.molbio.methds-reagnts was created by Paul Hengen
and is archived at:
<http://www-lecb.ncifcrf.gov/~pnh/FAQlist.html>
but is no longer updated. Paul has a host of other useful links at:
<http://www-fbsc.ncifcrf.gov/~pnh/>.

Paul Hengen explains his decision not to continue his TIBS column
at: <ftp://ftp.ncifcrf.gov/pub/methods/TIBS/sep97.txt>,
and a personal view about the purpose of Usenet by Russ Allbery can
be found at:
<http://home.xnet.com/~raven/Sysadmin/Rant.html>.

A demonstration of what the BIOSCI community is capable was written
up for the Biochemist and can be read online at:
<http://www.portlandpress.com/biochemist/cyber/0006/default.htm>,
with an addendum at:
<http://www2.mrc-lmb.cam.ac.uk/personal/rpg/CB/linux.html>.

12

The Roaming Scientist: Get Online, Manage Your E-mail and Exchange Files from Everywhere

Andrea Cabibbo

Contents

From: *The Internet for Cell and Molecular Biologists: Current Applications and Future Potential*
ISBN 1-898486-32-8 © 2002 Horizon Scientific Press, Wymondham, UK

Abstract

Science is an international business. Scientists often travel to other countries for variable periods of time and need to keep in touch and exchange material and information with their home lab and with collaborators worldwide. One of the most effective and simple ways to communicate and exchange documents, images, data and more general information is indeed e-mail. In most cases you will be able to use your e-mail account from all over the world, provided that the correct settings are entered in your e-mail application. E-mail has however some limitations as to the size of files that can be exchanged. Depending on the e-mail account, a variable limit on the size of attachments that can be sent and received exists. A limit also exists as to the total amount of megabites that can be stored in a personal mailbox on a mail server. This means that for the exchange of very large documents or very large amount of data, e-mail might be not well suited, and other systems have to be utilized, such as ftp, web sharing, the setting up of temporary simple web sites (see also chapter 4) or using an online storage facility.

In this chapter we will summarize the essential information required to read and send e-mail from everywhere (well, almost) and will provide some tips for the efficient exchange files of any (reasonable) size.

12.1 Getting online

Describing in detail the configuration of a computer for dial-up or ethernet connection to the internet is beyond the scope of this book. We will instead give some general suggestions on the various possibilities that exist to get connected to the net from different countries.

12.1.1 Host institution

If an institution abroad hosts you, most likely the LAN (Local Area Network) of the institution will be connected to the internet. Depending on the institution, you might either freely access computers or it might be necessary to ask for a personal account, that you will use to login on the available computers. This is often required in order for the institution to keep track of which user is performing specific tasks on the net.

If you want to connect your personal PowerBook to the net in a foreign/host institution on a regular basis, for instance for more than a couple of weeks, you should ask the system administrator to be assigned an personal IP number (see Chapter 1). The IP address should then be entered in the TCP/IP settings box of your PC, together with:

- the subnet mask

- the router address

- the DNS address (Domain Name Server, see Chapter 1).

Figure 12.1 provides an example of a configured TCP/IP control panel on a Macintosh.

Once your PC is correctly configured, you will be able to plug it to the ethernet LAN of the institution so as to be connected to the net.

Remember that, on a Mac, when you bring the PowerBook back home and connect with your modem though a dial-up connection, instead of the institution ethernet connection, you should switch back to the TCP/IP configuration of the ISP you connect to though the telephone line. You can configure several TCP/IP settings, for the different places from which you connect from, by using the "configurations" option of the TCP/IP control panel, which is accessed under the 'archive' menu of the control panel or by using the key combination 'apple'+ K (command + K).

In some instances, in order to surf the net, you might have to access the web through the 'proxy server' of the institution. The proxy sits between your computer and the net and acts as an intermediary: it passes your requests to the net and send the incoming web pages to

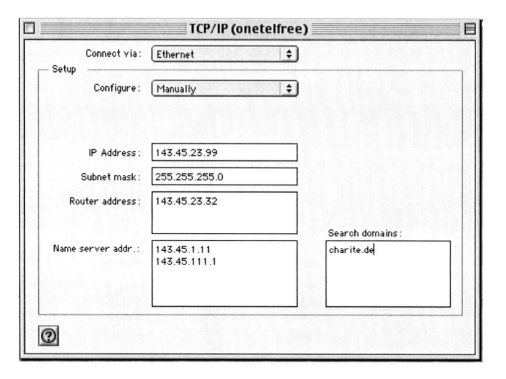

Figure 12.1. A configured TCP/IP control panel on a Macintosh.

your computer. This is an effective way of "masking" your computer from the external world. To connect through a proxy, you should modify the "preferences" or "settings" of your browser as shown in this example (Figure 12.2; Explorer 5.0 for Mac).

The precise location of the proxy settings depends on the browser and operating system that you are using.

12.1.2 Connect from home (Dial-Up)

If you want to connect with a dial-up (phone line and modem) connection from home in the foreign country, it can be a good idea to open an account on a free service provider before leaving for abroad, so that you will be able to connect immediately upon arrival at the destination. To find a free provider (ISP, Internet Service Provider) to

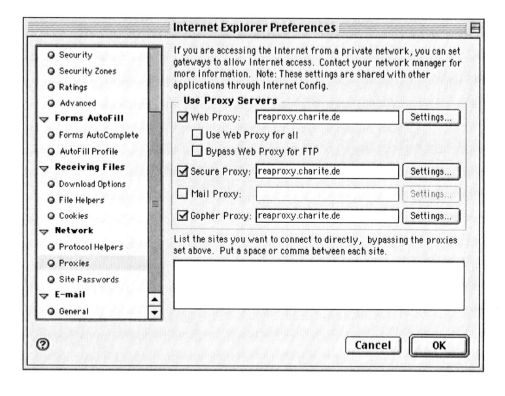

Figure 12.2. Location of the proxy settings in Internet Explorer.

register with in the destination country is quite a trivial task, unless (as we personally experienced recently) you are totally unable to understand the foreign language, which will also be used by the web sites of the foreign ISPs. In this case, an online translation service such as AltaVista Babelfish (see Chapter 13, http://babelfish.altavista.com) can be very valuable. For this procedure, we suggest the following steps:

(1)   Find a suitable foreign provider by using a search engine such as google (http://www.google.com) and typing something like 'Germany free internet access' or 'free isp Austria". If you are unable to read the foreign language, you can have the page translated. In Google there's a special feature for this (see Chapter 13). Be aware that automatic online translations can be very approximate and

are often exhilarating, but work well enough for this particular purpose.

(2)   Register with the ISP. This generally requires you to supply some personal details. You will then be provided with the data required to configure your computer for the dial-up connection. These data are:

•   A telephone number to dial in to
•   A username and a password.

    Sometimes you can select your personal username, sometimes not, depending on the ISP. You can nearly always select a password or change the default password that is supplied to you.

•   One or two DNS server addresses. The DNS server is a computer that will translate domain names to IP addresses (see Chapter 1), allowing you to surf the net.
•   An important piece of information that is not necessarily provided automatically, but that you should gather from your new ISP web site, is the address of the SMTP (Simple Mail Transfer Protocol) server of the ISP. As we will see in the next section, this is the server that allows you to send e-mail out as you are connected to a particular ISP. In most (but not all!) cases, the SMTP server will have an address such as mail.yourispname.com or smtp.yourispname.com. For instance, if you are connecting through the Tiscalinet ISP in Italy and you do not know the SMTP address, mail.tiscalinet.it and smtp.tiscalinet.it are two addresses worth trying. This setting should be entered in the proper place in your e-mail client. Details will be given in the next section.

(3)   Use the data supplied by the ISP to configure the new remote access connection that you will use abroad. The operations for this depend on your operating system. Many ISPs provide pages on their web site in which the

configuration is explained step by step. See for instance the instructions from the UK Compusolve ISP site at http://dial.csolve.co.uk/config.html. Note that the settings provided here are specific for Compusolve; you should use your ISP settings instead. Other good step-by-step tutorials for PC computers can be found at:
http://www.challenge.nm.org/ctg/login/dialup.shtml and
http://www.teckies.com/tutor/win111.html
while a Mac tutorial is at:
http://www.macinstruct.com/tutorials/modem/tcp/software.html

These steps should in principle allow you to step into your new room or apartment, plug in your telephone cable and get connected right away. Remember that telephone plugs can be different in different countries, so be sure to get a suitable cable from an electrical/computer store. Note that these operations are of course also valid to get connected to the net from your own country.

12.1.3 Internet cafes

Internet points or internet cafes are proliferating worldwide. It should not be difficult to find one near you. You can refer, for example, to the following web sites:

The Cybercafe Search Engine
http://www.cybercaptive.com/
When we visited, the site contained information on 7138 net-cafes in 167 different countries.

Internet Cafe Guide
http://www.netcafeguide.com
At the moment, 3500 cafes in 164 countries. A nice feature of this site is the netcafe map from which you can find cafes on earth by clicking an interactive map: http://www.netcafeguide.com/mapindex.htm.

To be more specific, you could try a Google search for something like 'netcafe australia' or 'cybercafe taiwan'.

In order to send and receive e-mail form an cybercafe you need to have a web-based e-mail account (see next section).

12.2 E-mail

There are basically two main ways to read your e-mail: through an e-mail client installed on your computer, such as Eudora (http://www.eudora.com) or Outlook Express (http://www.microsoft.com/office/outlook/), or directly on the web, through a web-based e-mail account. We will now briefly discuss both issues.

12.2.1 How to use your work e-mail account from home or from abroad

Your institution has assigned you a brand new e-mail address. You have configured your e-mail client with the correct setting and, in your institution, everything works fine. You now go home, connect your PowerBook through your home connection, and rapidly find out that while you are able to receive e-mails, you are apparently unable to send e-mails out, and receive error messages from your e-mail client as you try to do so. This is an extremely common situation. Why is that and how to solve this problem?

An e-mail application, such as Eudora or Outlook, will be "educated" to read your e-mail by entering, in the settings of the application, the following information:

- Your e-mail address. This will appear in the 'from' field of the e-mails you send out.

- Your username. This often, but not always, coincides with the part of your address preceding the @ symbol. If your address is john@YourIsp.com, your username is probably john.

- Your password

- The POP server address. If you connect through an ISP called yourisp, a typical pop address will be pop.yourisp.com. THE POP SERVER IS THE SERVER FROM WHICH YOU RETRIVE YOUR E-MAILS. Therefore, the POP server will not change as you change the location from which you connect (home, work...). Together with your username, the POP server defines the location of your personal e-mail account.

- The SMTP server. This is the bad guy, which is preventing from sending e-mails out. THE SMTP SERVER IS USED TO SEND E-MAILS OUT. Here's the golden rule to send out e-mail from any account, with any connection:

THE SMTP SERVER FIELD IN YOUR E-MAIL CLIENT MUST BE SET TO THE SMTP SERVER OF THE ISP/INSTITUTION THAT YOU ARE USING FOR THAT PARTICULAR CONNECTION TO THE INTERNET.

So if you are connecting from home, in order to use the work account to send mail out you should open the settings panel of your e-mail client and modify the address of the SMTP server to the one of the ISP that you use from home. The SMTP server address will often look something like mail.yourisp.com or smtp.youisp.com. When you go back to work, you should again modify the settings of the e-mail client in your PowerBook, so as to restore the address of your institutional SMTP server. If you do not have a PowerBook that you shuttle back and forth from home to work, but rather have two distinct computers at home and at work, the configuration of the e-mail client will be identical in all the fields BUT the SMTP server address, which will be the one of the institution in the work computer and the one of your ISP in the home computer.

This SMTP issue is the single most important piece of information that you should have very clear, in order to use your work e-mail form home or abroad and your personal home e-mail account

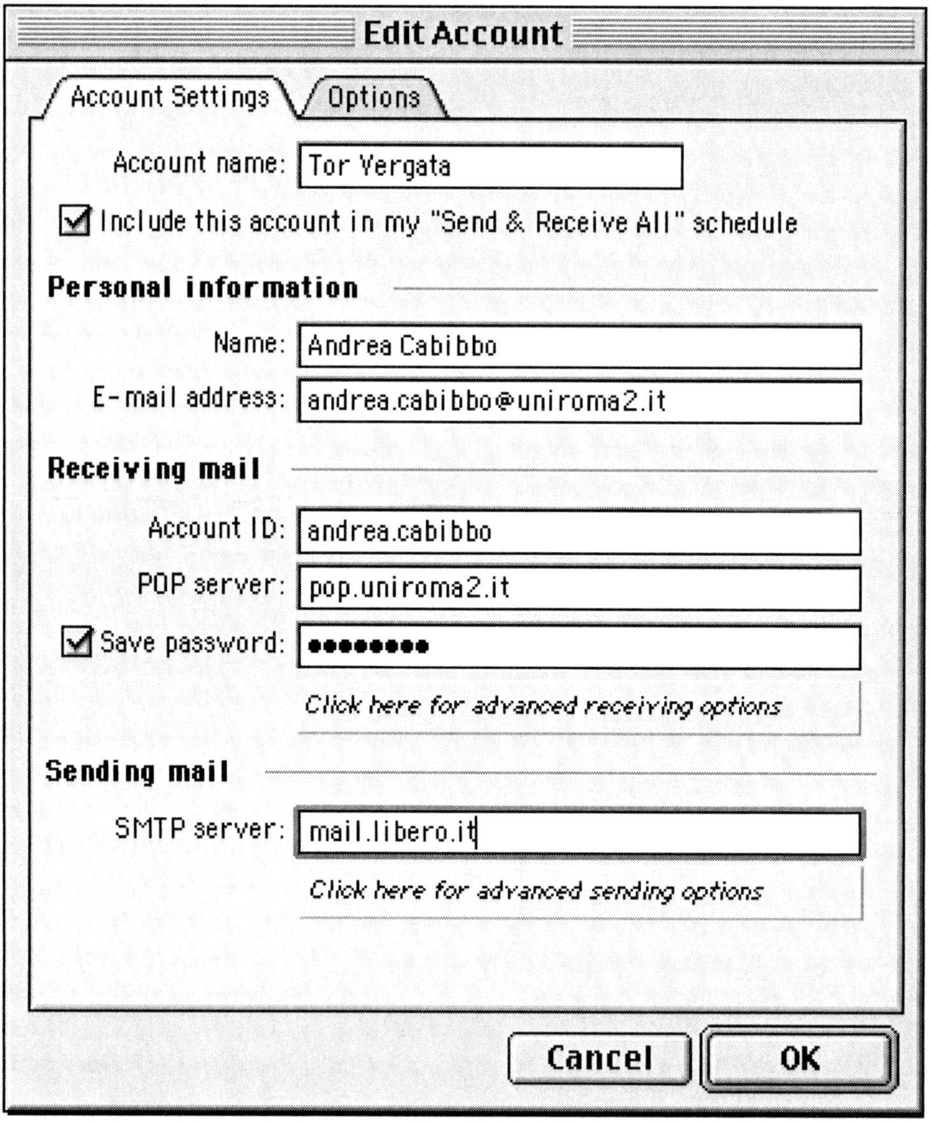

Edit Account

Account Settings \ / Options \

Account name: | Tor Vergata |

☑ Include this account in my "Send & Receive All" schedule

Personal information

Name: | Andrea Cabibbo |

E-mail address: | andrea.cabibbo@uniroma2.it |

Receiving mail

Account ID: | andrea.cabibbo |

POP server: | pop.uniroma2.it |

☑ Save password: | •••••••• |

Click here for advanced receiving options

Sending mail

SMTP server: | mail.libero.it |

Click here for advanced sending options

[Cancel]   [OK]

Figure 12.3.   An account at the second University of Rome "Tor Vergata" (http://www.uniroma2.it), configured in order to allow the sending of e-mails when connected from home through the "libero.it" ISP.

from work or from abroad. Figure 12.3 shows my work account at the second University of Rome "Tor Vergata" (http://www.uniroma2.it), configured in order to allow the sending of e-mails when connected from home through the "libero.it" ISP.

When connecting form the university, the SMTP has to be changed to mail.uniroma2.it.

If the LAN of your institution is sitting behind a firewall (see Chapter 3), this might not work for a number of reasons. In this case it is advisable to seek help from the system administrator.

12.2.2 Using a web-based e-mail account: read and send e-mail from any computer connected to the internet

Many web sites offer free web-based email accounts. These can be accessed through a web browser from any computer connected to the internet, including for instance computers in internet cafes all over the world (see 12.1.3). Moreover, for security reasons, it is strongly advised to never use your main e-mail address for any interaction or transaction on the net (see Chapter 3). So, even if you are not travelling, it is a good idea to set up a web-based free account.

All web-based accounts allow the user to send and receive e-mails, but they differ in a number of features that are summarized below:

Web based only or with POP3 capabilities.

Some web-based accounts can be accessed only via web browser. Others can be accessed both through the web and through a 'POP3' service, that is though e-mail clients such as Eudora and Outlook that are sitting on your computer. This is an important distinction and the choice very much depends on the kind of usage of the account.

For intensive usage, we strongly advise you to choose an account with POP3 capabilities. Many free accounts have a limit (~3 - 20 Mb) to the storage in your mailbox. Therefore you will be obliged to frequently delete some of your correspondence in order to keep the account active. POP3 capabilities allow you to efficiently download all your mail in the e-mail client of your computer and to have basically unlimited storage for your e-mail messages and attachments. The web access will allow you to manage your e-mail during trips or when, for any reason, you do not have access to your computer.

Excellent POP3 capabilities are offered for instance by the Yahoo mail service at http://mail.yahoo.com/. Unfortunately this service is no longer free.

POP3 capabilities are also offered by the "cellmail" service. This service provides you with an address of the kind yourname@cellmail.zzn.com and is accessible at http://cellmail.zzn.com/ or through the Bio-Web at http://cellbiol.com/.

For occasional usage or to exchange few e-mails, a web-based only account will be fine. To keep records of messages, you have the option of manually copying some important messages to a text file in your computer. A simple but reliable web-based only e-mail service can be found at http://www.xmail.com/.

Redirection

Some accounts offer redirection services. This means that all mails sent to this account will be forwarded to another account of your choice. This allows you to set a sort of 'façade' e-mail, while receiving all your messages at your main account, without necessarily giving out the address of this main account. This is a very effective way of fighting against spamming (see Chapter 3). If a particular e-mail address becomes clogged with unsolicited advertising messages, you can delete it without loosing your main e-mail account (which might be difficult to drop, especially if it is your work address).

A redirection service *de facto* adds POP3 capabilities to an account, because you can choose to redirect your e-mails to a POP3-based account.

Excellent redirection services are offered by the INAME service, at http://www.iname.com. Unfortunately this service is no longer free.

Collection

Some web-based accounts offer 'collection' services. This means that you will be able to read, from this account, mail sent to other POP3 based accounts of your choice (the ones you usually read through

Eudora or Outlook). This allows you, with some limitations, to transform your main, POP3 based account into a web account, or at least to read e-mail sent to your POP3 account from any computer, via a web browser. Yahoo mail offers collection services.

Address books, online agendas, filters, autoresponders, bells and whistles

Literally hundreds of web-based services exist. Just look at the result of a Google search for 'free web-based e-mail account' to get an idea. Each differs from the other for main features, such as POP3 capabilities, redirection, collection, and for a host of other subtleties. You should mainly look for a solid, reliable account, with as many of the above-described features as possible. All the services mentioned above are of good quality and have a number of features that you might find useful. Since the basic services are free, it is very easy to open an account and try the service for a while, in order to make a better choice.

Our personal recommendation is for the Yahoo service, which is very reliable and rich in advanced features that render the service very versatile and at the same time easy to use.

12.3 Some tips on file exchange

E-mail is indeed an effective way to exchange images, data and files. However, for sets of data larger than a few Mb, this becomes unpractical, because of the limit that many e-mail accounts impose on the size of attachments sent or received and the limit in the size of the mailboxes. In this section we suggest some possible solutions to the problem of file exchange.

12.3.1 FTP

The classical way of exchanging large sets of data is ftp (File Transfer Protocol). This was described in some detail in chapter 4. Basically, the computer containing the data will have an active 'ftp server' application running, which will be accessible through an 'ftp client' application that is sitting on the computer of the person who should retrieve the files. In practice, in order to share files with collaborators or with yourself, while travelling, you should ask your system administrator for access (address of the server, login and password) to an ftp server to which you will be able to login, through your favourite ftp client application (see Chapter 4), to upload and download files. By providing the login information to collaborators (permission should be asked of the system administrator), you will be able to share files with them. You will also be able to retrieve your files from any computer connected to the internet and with an ftp client installed.

If you have access to an ftp server, this is by far the most efficient solution for the exchange of files of any (reasonable) size. If not, don't worry and keep on reading.

12.3.2 Web site

It is very easy, as explained in chapter 4, to build-up a simple web page or web site. A possibility to share files is therefore to setup a web page that links to the files to be exchanged which you will upload to the web server together with the HTML page(s). Since the procedures to obtain web space from scientific institutions can sometimes be discouraging, you might consider the option of using one of the hundreds of free web hosting services existing on the net (consider for example the Tripod service at http://www.tripod.com/). These generally impose some kind of banners, advertising or pop-up on the pages of the subscribers, but this is not really a serious problem for the purpose of exchanging some files. Many ISPs provide, together with the internet access, a certain amount of web space, typically 20-50 Mb, which is optimal for this kind of usage. In simple terms the files will be uploaded to the web server, typically by ftp, together with a page that simply links to them. Everything should be put in the same folder. The HTML to make "data.doc", "article.rtf" and "images.zip" files downloadable is shown in Figure 12.4.

The browser automaticaly recognizes some file extensions, such as .doc, .rtf, .zip, .hqx, .bin and instead of trying to open the document in a browser window, offers to download the file.

```
<HTML>
<HEAD>
<TITLE>Data for Neil</TITLE>
</HEAD>
<BODY>
<P>
Here's the files Neil, just click to download.</P>
<P>
<A HREF="data.doc">Click here to download the data</A><BR>
<A HREF="article.rtf">Click here to download the article</A><BR>
<A HREF="images.zip">Click here to download the images</A><BR>
</P>
<P>
The images are not top quality but good enough for publication I guess. Let me know your opinion.<BR>
</P>
</BODY>
</HTML>
```

Figure 12.4. The HTML to make "data.doc", "article.rtf" and "images.zip" files downloadable.

If you did not understand a single word of this paragraph, you might want to go first through chapter 4 and then come back here.

12.3.3 Web sharing

Many applications exist that allow you to share entire folders of your PC or PowerBook on the web. Macintosh computers have an embedded web sharing control panel which is extremely simple to use. You select a folder to be shared, and start the sharing by pressing a button. The address of your shared folder on the web will be your IP address and will look something like http://192.195.62.33/. You can set the control panel so that it will ask for a username and password in order for visitors to access the content of your shared folder. You do not need to set up an HTML page that links to the files; sharing the folder will be enough, as the web sharing application will do the rest. Many similar applications exist for PCs. You can visit http://www.tucows.com/, select your operating system and search for web sharing or web server. You will find a wide range of freeware or shareware applications to choose from.

You should keep in mind that while at work you most likely have a static (fixed) IP address, every time that you connect from home with a dial-up connection, a new IP is assigned to your computer by the ISP (you have what is called a dynamic IP address). This means that if you want to give access to your data on the www to a collaborator while connecting from home, you should connect first, then activate the control panel, check your IP for this particular connection and communicate the IP to the collaborator. In this case, a web site might be a better choice as it provides a permanent address for the files. An essential advantage of web sharing is the possibility to password-protect your data. On the web site, your data are potentially available to everybody.

13

Bio-Bookmarks

Andrea Cabibbo and Manuela Helmer-Citterich

Contents

Abstract

Beyond the topics covered in the different chapters of this book, there are several other internet resources that can be of interest to biologists. In this chapter we shall try to give an overview of such resources, in order to complete the picture of the 'Bio-Web'. These and further links are available at http://cellbiol.com/. This list is by no means complete or exhaustive.

From: *The Internet for Cell and Molecular Biologists: Current Applications and Future Potential*
ISBN 1-898486-32-8 © 2002 Horizon Scientific Press, Wymondham, UK

13.1 Companies

One of the activities of the experimental biologist is to find and purchase reagents from companies. This is a very partial and incomplete listing of major suppliers of biological reagents.

Ambion
http://www.ambion.com/
'Ambion, the RNA company, develops innovative molecular biology research and diagnostic tools for the isolation, stabilization, synthesis, detection and quantitation of RNA'.

Amersham Biosciences
http://www.amershambiosciences.com/
Beyond providing a full range of research products for life sciences, Amersham is a world leader in the synthesis of radiolabelled chemicals.

ATCC, American Type Culture Collection
http://www.atcc.org/
Is one of the major repositories and supplier of bacterial strains, yeast and fungi strains, cell lines and much more.

BD Biosciences Clontech
http://www.clontech.com/
Supplies a full range of products for molecular biology research.

Biacore
http://www.biacore.com
A leader in the field of molecular interactions.

CRS Robotics
http://www.crsrobotics.com/
CRS designs and manufactures automated robotic genomic workstations for DNA purification and amplification as well as other genome-related assays. Clients include The Joint Genome Institute, The Institute for Genomic Research, The Australian Genome Project and several university labs.

EBS, Edimburgh Biocomputing Systems
http://www.mpsrch.com/
EBS develop software and systems to exploit the mass of biological data being created now from research results, such as those generated in the 'Human Genome Project'. These software products will give the potential to enhance the rapid progress of science and medicine in benefiting from such primary research projects, which generate huge 'libraries' of data including gene and protein sequences.

Eppendorf
http://www.eppendorf.com
'Eppendorf is a biotechnology company that develops, manufactures and distributes systems comprising instruments, consumables and reagents for use in laboratories worldwide. The company is focused on two business sectors:

• Bio Tools, which includes tools such as pipettes, dispensers and centrifuges, and consumables such as test tubes and pipette tips.

• Molecular Technologies, which includes instruments and systems for cell manipulation, automatic devices for high-throughput screening (HTS), complete systems for DNA multiplication, nucleic acid purification kits and biochips.'

Genentech
http://www.gene.com/
Leader in the field of recombinant DNA technology

Hamamatsu Photonics
http://www.hamamatsu.com/
The front edge in Bio-Imaging research

Immunology Consultants Laboratory, Inc
http://www.icllab.com/
Bulk producer of polyclonal antibodies and antigens. Human proteins, animal proteins, epitope tags, infectious diseases, conjugates and more.

Improvision
http://www.improvision.com
Develops software for scientific imaging. Products include Openlab for imaging of cell structure and function and Phylum for scientific image management.

Invitrogen-Gibco
http://www.invitrogen.com
Full range of molecular biology and cell culture reagents.

Jerini
http://www.jerini.de/
Specialist in field of high complexity protein, peptide and small molecules arrays.

Genset
http://www.gensetoligos.com/
Supplier of high quality oligonucleotides.

Interlab Cell Line Collection
http://www.iclc.it/
Stores and supplies cell lines from a huge collection.

MBI Fermentas
http://www.fermentas.com/
Molecular Biology products.

Millipore
http://www.millipore.com/
World leader in filtration technologies.

New England Biolabs
http://www.neb.com/
Molecular Biology products.

Perkin Elmer Life Sciences
http://lifesciences.perkinelmer.com/
Supplies products, services and technologies for functional genomics,

high throughput screening and drug discovery as well as for clinical screening. Products supplied comprise instrumentation, software and consumables, including reagents. One of the leaders in PCR products.

Promega
http://www.promega.com/
Molecular Biology products.

Qiagen
http://www.qiagen.com
Leader in DNA and RNA purification products.

Roche Molecular Biochemicals
http://biochem.roche.com/
Molecular biology and biochemistry products.

SciQuest
http://www.sciquest.com/
"SciQuest is a technology and solutions company that provides integrated e-commerce and research asset management solutions for research enterprises and their supply chain partners worldwide".

Sequitur
http://www.sequiturinc.com
Specialists in antisense technology.

Sigma-Aldrich
http://www.sigma-aldrich.com
From this page you will be able to access the full range of Fluka, Aldrich and Sigma products and chemicals.

Stratagene
http://www.stratagene.com/
Molecular Biology products.

The Antibody Resource Page
http://www.antibodyresource.com/
This is an excellent web page on antibodies: suppliers, custom

antibodies, educational resources on immunoglobulins and more Ab-related stuff.

13.2 Meetings

Here is a listing of sites where scientific meetings are announced. By no means complete.

Cold Spring Harbor Meetings
http://nucleus.cshl.org/meetings/

ELSO meetings
http://www.elso.org

EMBL Conferences, Courses and Workshops
http://www-db.embl-heidelberg.de/CoursesConferences.html

Euroconferences (Institut Pasteur)
http://www.pasteur.fr/applications/euroconf/

FASEB Meetings
http://www.faseb.org/meetings/

Genome and Biotechnology Meetings Calendar at ORNL
http://www.ornl.gov/TechResources/Human_Genome/CAL.HTML

Gordon Research Conferences Home Page
http://www.grc.uri.edu/

Hum-Molgen: Events in Bioscience and Medicine
http://www.hum-molgen.de/meetings/index.html

IBC Global Life Sciences
http://www.ibc-lifesci.com/

Keystone Symposia
http://www.symposia.com/

Knowledge Foundation Meetings Listing
http://www.knowledgefoundation.com/

Meetings page at Biolinks.com
http://www.biolinks.com/
Follow the 'meetings' link from the home page.

13.3 Laboratory Protocols

Darmouth Protocols
http://www.dartmouth.edu/artsci/bio/ambros/protocols/molbio.html

GOAA Protocols
http://research.nwfsc.noaa.gov/protocols.html

Material and Methods Online
http://www.methods-online.net/ or http://www.methods.info

Protocols at the RBB Department of Genetics
http://www.biol.rug.nl/lacto/protocols/protocols.html

Protocols at Highveld.com
http://www.highveld.com/protocols.html

Protocol Online
http://www.protocol-online.net/
Very good! Subdivided by arguments: DNA, RNA, Protein, DNA-Protein interaction, Organelles, Carbohydrates, Bacteria, Yeast, PCR, Cell culture....

Protocols for recombinant DNA isolation, cloning and sequencing
http://www.genome.ou.edu/protocol_book/protocol_index.html
Web version of "DNA Isolation and Sequencing" (Essential Techniques Series) by Bruce A. Roe, Judy S. Crabtree and Akbar S. Khan Published by John Wiley and Sons.

13.4 Biological directories and sites

Many sites exist that list biological resources on the net. Here's a short selection. A couple of important biological sites (not directories) were also included here.

The About.com Biology Section
http://biology.about.com/
Excellent directory covering all branches of modern biology. Maintained by Regina Bailey.

All the Virology on the WWW
http://www.virology.net/
Excellent site that attempts to cover all the virology related information available on the web. Maintained by David M. Sander.

Art's Biotech Resource
http://www.ahpcc.unm.edu/~aroberts/index.htm
Biochemistry, Biophysics, Molecular Biology and Bioinformatics. Really cool! Maintained by Arthur G. Roberts.

Bioinformatik
http://www.bioinformatik.de/
Bioinformatics links and news.

The Bio-Web
http://cellbiol.com
Online resources for molecular and cell biologists: sequence analysis tools, journals, companies, meetings, news, software, books and more stuff. Maintained by Andrea Cabibbo.

Cell and Molecular Biology Online
http://www.celbio.com
An informational resource for cell and molecular biologists. Top class. Maintained by Pamela Gannon.

Highveld.com
http://www.highveld.com
Exhaustive resource for links and information on molecular and cell biology on the net. Hosts, among other web sites, the very popular 'PCR Jump Station', that can be accessed at:
http://www.highveld.com/pcr.html

Molecular Biology Shortcuts
http://www.mbshortcuts.com
Excellent collection of resources for Molecular Biologists including a local sequence analysis tools page, a nice bookstore and more interesting stuff.

The Open Directory Project (DMOZ)Biology section
http://dmoz.org/Science/Biology/

Pedro's Research Tools: Selection of web sites for molecular biology
http://www.public.iastate.edu/~pedro/research_tools.html
In this site you will find a list of useful sites, with a very short introduction. This site has been extremely popular, as it was basically the first bio-directory on the web. It is now a bit out of date.

Science Advisory Board
http://www.scienceboard.net/
A science 'infomediary' company that aims to use the power of consumers to obtain better deals from life science suppliers. Membership is open to clinical and academic scientists and investigators. Members take part in surveys, and are rewarded for their time in a variety of ways (usually Amazon gift certificates and a redeemable 'points' system). There is a forum open to all members and a resources section with book reviews, product literature, meeting announcements, etc.

The WWW Virtual Library: Cell Biology Section
http://vlib.org/Science/Cell_Biology/
Outstanding. Maintained by Gabriel Fenteany.

Yahoo Biology section
http://dir.yahoo.com/Science/Biology/

13.5 Microarray resources and databases

Microarray informatics at EBI
http://www.ebi.ac.uk/microarray/
A web-based tool for microarray data analysis and a public repository
(ArrayExpress) for microarray based gene expression data.

Yeast and E.coli RNA expression database
http://twod.med.harvard.edu/ExpressDB/
ExpressDB is a relational database containing yeast and *E.coli* RNA
expression data.

Gene Expression Project
http://www.ncgr.org/genex/
GeneX provides an internet-available repository of gene expression
data with an integrated tool-set for data analysis and comparison.

Gene Expression Omnibus
http://www.ncbi.nlm.nih.gov/geo/
GEO, located at NCBI, is a gene expression and hybridization array
data repository, as well as an online resource for the retrieval of gene
expression data from any organism or artificial source.

Stanford Microarray Database
http://genome-www5.stanford.edu/MicroArray/SMD/
SMD stores raw and normalized data from microarray experiments,
as well as their corresponding image files in a public repository for a
list of organisms. In addition, SMD provides web tools for data
retrieval, analysis and visualization.

13.6 Protein interaction resources

A short selection of databases and methods dedicated to protein-protein interaction.

Receptor-Ligand database
http://relibase.ebi.ac.uk/reli-cgi/rll?/reli-cgi/general_layout.pl+home
A database and software dedicated to protein-ligand interaction.

Molecular Interaction Database
http://cbm.bio.uniroma2.it/mint/
MINT is a relational database storing protein-protein interactions extracted from the literature by expert curators. Beyond cataloguing the formation of binary complexes, MINT catalogues other type of functional interactions namely enzymatic modifications of one of the partners. Both direct and indirect relationships are considered. Nice graphical networks provided.

Biomolecular Interaction Network Database
http://www.bind.ca/
BIND is a database storing full descriptions of interactions, molecular complexes and pathways. It can be used to study networks of interactions, to map pathways across taxonomic branches and to generate information for kinetic simulations.

Database of Interacting Proteins
http://dip.doe-mbi.ucla.edu/
DIP catalogs experimentally determined interactions between proteins, but considers also computationally inferred interactions. From the David Eisenberg group.

Protein-Protein Interaction database
http://pronet.doubletwist.com/
ProNet contains information on human protein-protein interactions described in the literature. At the time of writing only interactions identified with the two hybrid system are included.

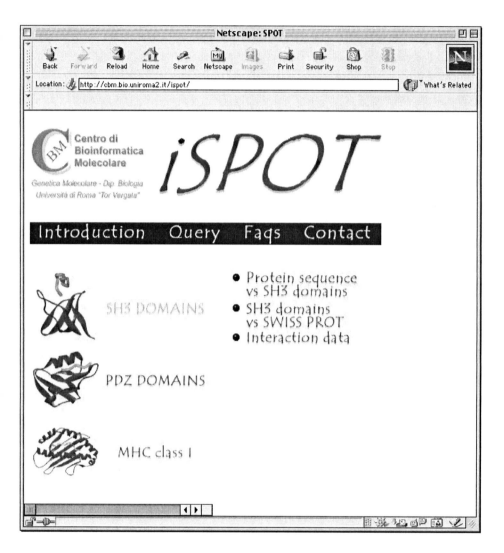

Figure 13.1. iSPOT (Sequence Prediction Of Target) is a web tool developed for the prediction of protein-protein interaction mediated by families of peptide recognition modules.

Prediction of interaction specificity between protein modules
http://cbm.bio.uniroma2.it/ispot/
iSPOT (Sequence Prediction Of Target) is a web tool developed for the prediction of protein-protein interaction mediated by families of peptide recognition modules (Figure 13.1). At present it works on SH3 and PDZ domains.

Prediction of inter-residue contacts
http://prion.biocomp.unibo.it/cornet.html
The method is a neural network based predictor that uses as input correlated mutations, sequence conservation, predicted secondary structure and evolutionary information.

Prediction of contacts among residues
http://montblanc.cnb.uam.es/cnb_pred/
PDGCON, CONcons and CONhydro are virtual servers for contact predictions.

13.7 Useful sites for lessons and presentations

All over the world people are preparing lessons and presentations and a lot of work is duplicated probably many times. Some very high quality didactic material is freely available online. Don't miss it!

http://www.ergito.com/
This site is a fantastic place: it offers the online version of GENES. It is based upon the same material as the print edition of GENES, but is continuously updated. The most recent print version of GENES is the 7th edition, published in January 2000, while this site was updated in October 2001. The text, figures and references are all freely available after registration. A subsession is dedicated to 'Great experiments' (e.g. 'Isolation of repressor' by Mark Ptashne). Excellent support for teaching.

http://www.oup.co.uk/best.textbooks/biochemistry/genesvii/illustrations/
This site offers high quality pictures (from well-known and used textbooks) for presentations and teaching. The figures can be downloaded and, if necessary, even modified on your PC. Enrich with good pictures the impact of your course!

http://www.cellsalive.com/menu.htm
A very interesting site offering animations and movies describing fundamental topics of molecular and cell biology (cell cycle, mitosis and others, step by step). Great fun and very useful material for lessons!

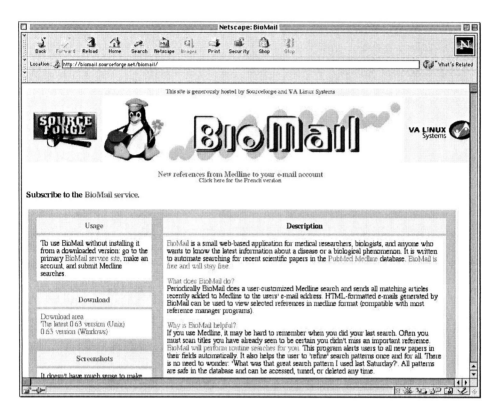

Figure 13.2. BioMail: A free web-based application for medical researchers and biologists.

13.8 Biology servers

PubMed
http://www.ncbi.nlm.nih.gov/entrez/
An absolute MUST. Free Medline searches (it works with simple keywords or with author names) providing abstracts, literature citations and linking to full-text journal articles at web sites of participating publishers.

BioMail Service
http://biomail.sourceforge.net/biomail/
BioMail is a free web-based application for medical researchers and biologists. It is written to automate searching for recent scientific papers

in the PubMed Medline database. Periodically BioMail does a user-customized Medline search and sends all matching articles recently added to Medline to the users' e-mail address. HTML-formatted e-mails generated by BioMail can be used to view selected references in medline format (compatible with most reference manager programs, Figure 13.2).

The CMS Molecular Biology Resource
http://restools.sdsc.edu/
This website is a compendium of electronic and Internet-accessible tools and resources for Molecular Biology, Biotechnology, Molecular Evolution, Biochemistry, and Biomolecular Modeling. Very good!

ExPASy Molecular Biology Server
http://www.expasy.ch/
The ExPASy (Expert Protein Analysis System) proteomics server of the Swiss Institute of Bioinformatics (SIB) is dedicated to the analysis of protein sequences and structures as well as 2-D PAGE. It offers links to many services and databases with a special focus on proteomics. A must.

RNA structure Database
http://www.rnabase.org/
RNABase stores all RNA containing structures from both the NDB and RCSB. In the near future it will incorporate a number of automatic conformational analyses.

The RNA world website
http://www.imb-jena.de/RNA.html
A useful source of information and links for people interested in RNA analysis, sequences, secondary structures, 3D structures. Data and web tools.

Sequence Retrieval System
http://srs.embl-heidelberg.de:8000/srs5/
http://srs.ebi.ac.uk/
A sequence retrieval system which accesses different biological databases and is able to give information about different interconnected

biological databases. Very useful, but not so user-friendly…

Free software for handling sequences
http://www.hgmp.mrc.ac.uk/Software/EMBOSS/
http://www2.no.embnet.org/Pise/index.php3
EMBOSS is a package of high-quality FREE Open Source software
for sequence analysis. It can be downloaded from the first proposed
address. The same package is also available through the web at the
Norway EMBnet site. Results are given on the web or by email. Gold
medal site…

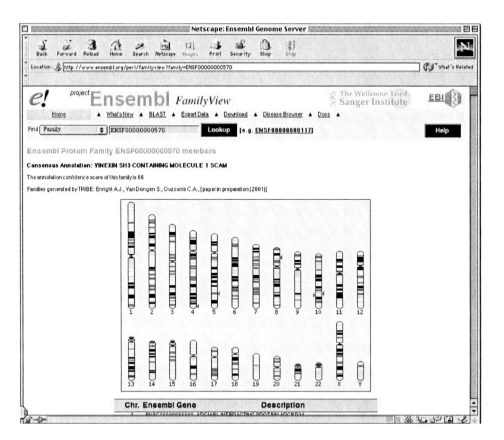

Figure 13.3. Ensembl: A very useful site for people interested in human genome.

Annotation and tools for human genome research
http://www.ensembl.org/
A very useful site for people interested in human genome. You can
search the human genome for similarity to a query sequence. You can
also search the genome for keywords: in this case all the annotations
are searched and you get a nice graphic overview of the known genes
associated to the given keyword (Figure 13.3)

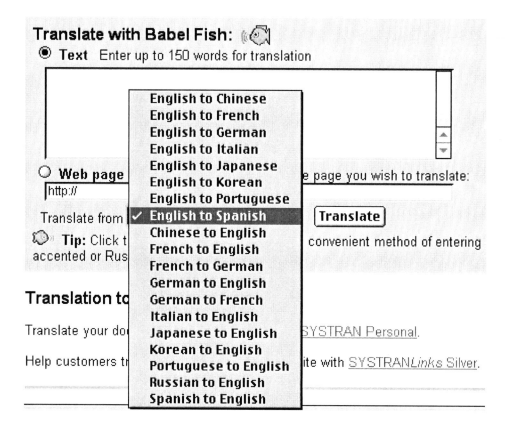

Figure 13.4. Online translations by Altavista (Babelfish, http://babelfish.altavista.com).

13.9 Miscellanea

Very useful links for biologists are often unrelated to biology. Here's a short list.

Webster dictionary online
http://work.ucsd.edu:5141/cgi-bin/http_webster
The Webster online. A very useful feature is the possibility of looking not only for a perfect match to your query but also for an approximate match. This allows searching for words for which the exact spelling is not known. For this reason this resource is great for spell check while writing an article.

Oxford English Dictionary online
http://www.oed.com/
The one true dictionary. Access is free to academic institutions.

Dictionary online
http://www.dictionary.com
Another online dictionary. Offers the possibility to translate text or websites at http://translator.dictionary.com/. This feature is very similar to the Altavista babelfish site.

Online translations by Altavista (Babelfish)
http://babelfish.altavista.com
You can translate text or websites. While you navigate websites in babelfish, every link that you follow is also translated (Figure 13.4).
      The translation is far from perfect and often very funny, however it gives you an idea, for instance, of what this Japanese site is about, without having any knowledge of Japanese.

Website translation by Google
http://www.google.com
When you perform a search with google, you are offered the possibility to translate any of the results to the languange of the google site that you are using. For instance, the www.google.com will translate into English, while the www.google.it site will translate into Italian. The 'translate this page' link is located near the title of the web site (Figure 13.5).

Breve storia della carta - [Translate this page]
... dalla pergamena. La pergamena fu il **supporto di**
scrittura dal III secolo dC ... Carta
ecologica è la carta, **di cellulosa** o riciclata, per la cui ...
www.cartaecartone.it/breve.htm - 9k - Cached - Similar
pages

Figure 13.5. Website translation by Google.

Software download site
http://www.twocows.com/
Tucows provides software on a "freeware" or "shareware" basis through a network of more than 1000 partner sites. A search engine is accessible to obtain a list of available programs (with short description and evaluation) for all operating systems.

Maps
Several web sites offer the possibility to generate maps, often very detailed, of locations/towns of interest.

MapBlast is at http://www.mapblast.com. Requires a free registration and covers virtually all the earth. Excellent service.

Multimap is at http://www.multimap.com/. You start with a map of the earth and have the possibility of repeatedly zooming in by clicking a selected location. Very detailed output, easy to use.

Yahoo currency converter
http://finance.yahoo.com/m3?u
Imput a value in any currency and have it converted to any other currency of your choice. Very simple, useful and reliable (Figure 13.6).

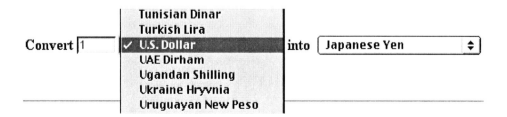

Figure 13.6. Yahoo currency converter.

Yahoo weather
http://weather.yahoo.com/
Preview weather in all the world.

Trains timetables
Swiss Railroad Timetable is at http://www.sbb.ch/pv/. Queries for all Europe. Versions in German, French, Italian, English.

TheTrainLine is at http://www.thetrainline.com/. Gives the opportunity to find times, book tickets or reserve seats for any train operator in mainland UK.

Index

Books of Related Interest

Gene Cloning and Analysis: Current Innovations

Genetic Engineering with PCR

An Introduction to Molecular Biology

Probiotics: A Critical Review

Prions: Molecular and Cellular Biology

Peptide Nucleic Acids: Protocols and Applications

Intracellular Ribozyme Applications: Principles and Protocols

Prokaryotic Nitrogen Fixation: A Model System for the Analysis
    of a Biological Process

Molecular Marine Microbiology

NMR in Microbiology: Theory and Applications

Oral Bacterial Ecology: the Molecular Basis

Development of Novel Antimicrobial Agents: Emerging Strategies

Cold Shock Response and Adaptation

Flow Cytometry for Research Scientists: Principles and Applications

Helicobacter pylori: Molecular and Cellular Biology

The Spirochetes: Molecular and Cellular Biology

Environmental Molecular Microbiology: Protocols and Applications

Advanced Topics in Molecular Biology

Industrial and Environmental Biotechnology

Genomes and Databases on the Internet

Microbial Multidrug Efflux

Emerging Strategies in the Fight Against Meningitis:
    Molecular and Cellular Aspects

The Bacterial Phosphotransferase System

For further information on these books contact:

Horizon Scientific Press, 32 Hewitts Ln, Wymondham, Norfolk, NR18 0JA, U.K.
Tel: +44(0)1953-601106. Fax: +44(0)1953-603068. rab@horizonpress.com

**Our Web site has details of all our books including full chapter
abstracts, book reviews, and ordering information:**

www.horizonpress.com